112 Combinatorial Problems
from the AwesomeMath Summer Program

112 Combinatorial Problems
from the AwesomeMath Summer Program

Vlad Matei

Elizabeth Reiland

Library of Congress Control Number: 2016915802

ISBN-10: 0-9968745-2-6
ISBN-13: 978-0-9968745-2-6

© 2016 XYZ Press, LLC

All rights reserved. This work may not be translated or copied in whole or in part without the written permission of the publisher (XYZ Press, LLC, 3425 Neiman Rd., Plano, TX 75025, USA) and the authors except for brief excerpts in connection with reviews or scholarly analysis. Use in connection with any form of information storage and retrieval, electronic adaptation, computer software, or by similar or dissimilar methodology now known or hereafter developed is forbidden. The use in this publication of tradenames, trademarks, service marks and similar terms, even if they are not identified as such, is not to be taken as an expression of opinion as to whether or not they are subject to proprietary rights.

9 8 7 6 5 4 3 2 1

www.awesomemath.org

Cover design by Iury Ulzutuev

Preface

Combinatorics is a fascinating branch of mathematics centered around counting various objects and sets. Counting problems make regular appearances on middle and high school mathematics competitions despite the fact that combinatorics is generally covered only very briefly in high school math courses. This is not, however, because combinatorics requires higher level math as a prerequisite; indeed, many counting problems are accessible to anyone with a solid background in arithmetic and some basic algebra.

This book gives students a chance to explore some introductory to intermediate topics in combinatorics. We include chapters featuring tools for solving counting problems, proof techniques, and more to give students a broad foundation to build on. It is worth noting that some sections of this book are significantly more challenging than others. In particular, the chapters on *Invariants, Counting in more than one way*, and *Generating functions* cover topics that are considered fairly advanced; readers should not be discouraged if they do not immediately grasp these concepts. Though counting problems in particular are accessible to anyone, that does not mean they are trivial. One of the trickiest aspects of solving a counting problem is determining which tool or trick should be used. To help readers become accustomed to dealing with these subtleties, each section includes several example problems of varying difficulty with solutions to demonstrate how the different techniques may be applied in practice.

Following these topic-based segments we have included several introductory and advanced problems for students to tackle by themselves. These were carefully selected to enable the reader to further hone their problem solving skills based on the material presented in the chapters. Students can check their work in the final part of this book, which includes detailed solutions to these problems.

Several of the problems that appear in this book are pulled from various mathematics competitions worldwide. We would like to express our gratitude to the many writers who have contributed to these contests and provided us with such a rich selection of exercises. We would also like to thank Dr. Titu Andreescu for giving us the opportunity and encouragement to write this book and Dr. Richard Stong, Dr. Branislav Kisacanin, and Dr. Walter Stromquist for their thoughtful feedback, which helped us shape this text to be the absolute best it could be.

We hope you enjoy the problems!

Elizabeth "Lizard" Reiland
Vlad Matei

Contents

Preface	v
1 Counting Basics	1
2 Permutations and Combinations	9
3 Stars and Bars and Multinomials	19
4 Principle of Inclusion-Exclusion	31
5 Pascal's Triangle and the Binomial Theorem	41
6 Counting in More Than One Way	51
7 Pigeonhole Principle	61
8 Induction	67
9 Recurrence Relations	79
10 Graph Theory	91
11 Invariants	99
12 Combinatorial Geometry	109
13 Generating Functions	121
14 Probabilities and Probabilistic Method	137
15 Introductory Problems	147
16 Advanced Problems	155

17 Solutions for Introductory Problems	**163**
18 Solutions for Advanced Problems	**195**
Appendix: Recurrence Relations	**237**
Glossary	**241**
Other Books from XYZ Press	**245**

Chapter 1

Counting Basics

Before we jump into counting, we will go over some set theory definitions and notation that is important to our study of combinatorics. These are common terms that appear throughout the mathematical literature, so it is good to learn and remember them.

Definition 1. A *set* is a collection of distinct elements whose order is not important. We can specify a set by listing its elements such as $\{1, 2, 4, 8, 16\}$ or $\{3, 5, 7, \ldots, 19\}$. Notice that our definition means that, for example, $\{1, 2, 4\}$, $\{2, 4, 1\}$, and even $\{1, 1, 2, 2, 4\}$ are exactly the same set.

We can also use *set builder* notation where we specify a condition used to determine which elements belong to the set such as $\{x \mid 1 < x < 17, x \text{ is an integer}\}$. The bar $\mid$ can be read as "such that," so this set is all values x such that x is an integer and $1 < x < 17$. Thus this set is simply $\{2, 3, \ldots, 16\}$. Another example of set builder notation is $\{(x, y) \mid x \text{ and } y \text{ are real numbers}, y = 3x + 4\}$. Note that this set contains an infinite number of ordered pairs (x, y).

- The *empty set* is the set which contains no elements. We denote it as $\{\ \}$ or $\emptyset$.

- The notation $x \in A$ (read "x is in A" or "x is an element of A") means that the element x is included in the set A. We use the notation $y \notin A$ (read "y is not in A" or "y is not an element of A") to indicate that y is not included in the set A.

- We say that a set A is a *subset* of a set B (denoted $A \subseteq B$) if every element of A is an element of B (i.e., $x \in A$ implies $x \in B$).

- Two sets A and B are *equal* (denoted $A = B$) if they contain exactly the same elements. (One common way to prove $A = B$ is to show that $A \subseteq B$ and $B \subseteq A$. Keep this in mind!)

- The *union* of two sets A and B (denoted $A \cup B$) is the set of all elements in either A or B: $\{x \mid x \in A \text{ or } x \in B\}$. The *intersection* of A and B (denoted $A \cap B$) is the set of all elements belonging to both A and B: $\{x \mid x \in A \text{ and } x \in B\}$. These definitions can be extended to more than two sets in the intuitive way:

$$S_1 \cup S_2 \cup \cdots \cup S_k = \{x \mid x \in S_i \text{ for some } i, 1 \leq i \leq k\}$$

$$S_1 \cap S_2 \cap \cdots \cap S_k = \{x \mid x \in S_i \text{ for all } i, 1 \leq i \leq k\}$$

- We say two sets A and B are *disjoint* if they have no elements in common (i.e., if $A \cap B = \emptyset$).

- The *set difference* of the set A and the set B (denoted $A \backslash B$) is the set of elements that are in A but not in B. This notation is used even when B is not a subset of A; for example, $\{1, 2, 3\} \setminus \{3, 4\}$ is $\{1, 2\}$.

- If we have a *universal set* U which contains all of the objects we are interested in, we can define the *complement* of a set A (denoted A^c) as the collection of elements not in A (i.e., $A^c = U \backslash A$). For example, if we are working with the set of integers, the complement of the set of even numbers would be the set of odd numbers. (*Note:* we have to have some universal set in order for the idea of a complement to make sense!)

- The *cardinality* or *size* of a set A (denoted $|A|$) is the number of elements in that set.

Though these definitions may seem straightforward, there are some surprisingly subtle issues in set theory. It is possible for elements of a set to be sets in their own right. For example, one could take the set A of all subsets of $\{1, 2, 3\}$ (called the *power set* or A). We have

$$A = \{\emptyset, \{1\}, \{2\}, \{3\}, \{1, 2\}, \{1, 3\}, \{2, 3\}, \{1, 2, 3\}\}.$$

The elements of A are sets. One could iterate this idea to build sets whose elements are sets of sets, and so on. Another interesting example is the set $B = \{\emptyset\}$. Notice that B is not the empty set, but rather the set containing the empty set. The size of the empty set is $|\emptyset| = 0$, but we have $|B| = 1$.

One might then worry about whether a set A could contain itself as an element, $A \in A$. To avoid this one might try to restrict to the set of all sets that do not contain themselves, $B = \{A : A \notin A\}$. Thinking about whether B contains itself will lead you to what is known as Russell's paradox. These issues can be fun but will not be relevant to this text, since our sets will be explicitly defined and usually finite.

Counting Basics

As we start thinking about counting, there are two essential rules that will show up in almost every problem you encounter. Once you've done a bit of counting, you'll find yourself using these without even thinking about them. We will state these principles formally in a moment, but first we will examine a simple example.

Example 1. Suppose we are at a clothing store which offers 16 different shirts, 9 different pairs of pants, and 3 different pairs of shoes. How many ways are there to purchase an article of clothing?

Before we discuss the solution to this exercise, note that Example 1 illustrates an important fact about combinatorics problems. It is more fun to phrase combinatorics problems in simple English, and this is the way you will often see them. However, English is not as precise a language as mathematics, and we generally do not want to include long lists of disclaimers and explanations to make the problems technically precise since this would defeat the point of using simple English.

One of the first steps you should take when approaching a combinatorics problem is to decide how you want to interpret the English. For instance, in solving Example 1, we will implicitly assume that the only types of articles of clothing are the three mentioned (shirts, pants, and shoes) and that shoes have to be purchased in a pair. Mathematicians generally agree on how to interpret problems, and this is one of the things you will pick up going through the examples. If you are uncertain how to interpret a problem statement and are unable to ask someone to clarify, make your best attempt at an appropriate interpretation and be sure to note the assumptions you have made in your solution.

Having made these notes, let us now solve Example 1.

Solution. Because an article of clothing is either a shirt, a pair of pants, or a pair of shoes we can simply add up the number of each type of clothes to find $16 + 9 + 3 = 28$ possible ways to buy an article of clothing. □

The counting in this exercise was fairly straightforward, but it illustrates an application of the Sum Rule, a generalized principle which can be used to solve much more complicated problems. The formal statement of the Sum Rule is as follows:

Theorem 1. (Sum Rule) *If $A_1, A_2, \ldots, A_n$ are pairwise disjoint sets (i.e., if no pair of sets have elements in common), then*

$$|A_1 \cup A_2 \cup \cdots \cup A_n| = |A_1| + |A_2| + \cdots + |A_n|.$$

While this may seem like a lot of fancy notation, in practice this rule just tells us that if we are counting the possible ways to pick an object from one of several different sets that do not overlap, we just need to add up the sizes of the individual sets. If the sets do overlap, we will need to be a bit more careful; we discuss how to deal with this possibility in the inclusion exclusion section. Let us give a simple example of applying this principle

Example 2. Let $X = \{1, 2, \ldots, 200\}$. We define

$$S = \{(a, b, c) \mid a, b, c \in X, \ a < b \text{ and } a < c\}.$$

How many elements does S have?

Solution. Note that we can split S up into disjoint set A_k where k is the value of a and $1 \leq k \leq 199$. Note that since $b > k$ and $c > k$ we have $200 - k$ choices for b and $200 - k$ choices for c. Thus $|A_k| = (200 - k)^2$. Using the addition principle we obtain $|S|$. □

Example 3. Suppose we are at a clothing store which offers 16 different shirts and 9 different pairs of pants. How many ways are there to purchase an outfit consisting of one shirt and one pair of pants?

Solution. To help facilitate our counting, let us build a table. Each row of the table will represent a particular shirt, whereas each column will represent a particular pair of pants. A particular cell in the table will correspond to the outfit consisting of the shirt indicated by the row and the pair of pants indicated by the column of that cell. Since each cell will represent one distinct outfit, and every outfit appears in exactly one cell, our number of outfits is simply equal to the number of cells in our table. Since we have 16 shirts and 9 pairs of pants, there are $16 \cdot 9 = 144$ cells in our table, and thus 144 possible outfits we could buy. □

Notice that if we wanted to create an oufit consisting of a shirt *and* a pair of pants *and* a pair of shoes, we could expand on this idea to make a three dimensional table with one coordinate representing shirts, a second representing pants, and the last representing shoes. Similarly, if we had n selections to make, we could imagine counting cells in an n-dimensional table. This brings us to our other basic rule:

Theorem 2. (Product Rule) *If we have a sequence of n choices to make with X_1 possibilities for the first choice, X_2 possibilities for the second choice, and so on up to X_n choices for the nth choice, there are a total of $X_1 \cdot X_2 \cdot \ldots \cdot X_n$ ways to make our choices.*

For most problems we will apply both the Sum Rule and Product Rule to get us to our final solution. By using the Sum Rule we can break problems into a collection of cases where each case is relatively simple to count (generally by employing the Product Rule) as illustrated in the next example.

Example 4. How many three-digit numbers have exactly one even digit?

Solution. We will look at three different cases here: the case where the first digit is even and the other two are odd, the case where the middle digit is even and the other two are odd, and the case where the last digit is even and the other two are odd. Since these cases do not overlap, we can count each individually, then apply the Sum Rule to get our final answer.

We can think of creating a three-digit number as a series of three steps: choosing the first digit, choosing the second digit, and choosing the final digit. In the case where the first digit is even and the other two are odd, there are 4 choices for the first digit $(2, 4, 6, 8)$ since it must be even and cannot be zero (otherwise we would not have a three-digit number). Since the second and third digits are both odd, there are 5 possibilities for each $(1, 3, 5, 7, 9)$. Thus the Product Rule tells us there are $4 \cdot 5 \cdot 5$ three-digit numbers fitting this case. In the case where the middle digit is even and the other two are odd, every digit has 5 possibilities: $1, 3, 5, 7, 9$ for the odd digits and $0, 2, 4, 6, 8$ for the even digit. Overall then, there are $5 \cdot 5 \cdot 5$ three-digit numbers in this case. Similarly, there are $5 \cdot 5 \cdot 5$ numbers satisfying the case where the last digit is even and the other two are odd. Putting these three cases together using the Sum Rule, we have $4 \cdot 5 \cdot 5 + 5 \cdot 5 \cdot 5 + 5 \cdot 5 \cdot 5 = 350$ three-digit numbers with exactly one even digit. $\square$

One more basic but very useful technique to keep in mind is *complementary counting*. Suppose we are interested in determining the size of a set A. If we have a finite universal set U, we know by the Sum Rule that $|A| + |A^c| = |U|$. Rearranging, we find $|A| = |U| - |A^c|$. We can take advantage of this to help us determine the size of A. In particular, we can determine the size of our universal set and the size of the complement of A, then subtract. In some cases this may be significantly easier than trying to directly count A. If you see the words "at least" in a problem, complementary counting will often be a good method to consider.

Example 5. How many four-digit positive integers have at least one digits that is a 2 or a 3?

(2006 AMC 10A)

Solution. Let's first count the total number of four-digit positive integers. The first digit must be from 1 to 9, so we have 9 choices. For each of the three remaining digits, we need a value from 0 to 9 so there are 10 choices each.

Thus in total there are $9 \cdot 10 \cdot 10 \cdot 10 = 9 \cdot 10^3 = 9000$ four-digit positive integers.

Next we count how many four-digit positive integers DO NOT contain a 2 or a 3. Then we have 7 choices for our first digit (1,4,5,6,7,8, or 9) and 8 for the remaining three-digits. This gives a total of $7 \cdot 8^3$ four-digit integers not containing a 2 or a 3. Subtracting this from our total, we conclude that there are $9000 - 7 \cdot 8^3 = 5416$ four-digit positive integers that have at least one digit that is a 2 or a 3. $\square$

Let's look at some examples of problems making use of the techniques we've learned thus far.

Example 6. How many subsets of $\{1, 2, \ldots, n\}$ are there? (Note: This quantity will come up frequently in problems, so it's a useful fact to remember.)

Solution. Consider an element i ($1 \leq i \leq n$). As we construct a subset S, we have two choices for i: Either it is in S or it is not in S. Since we must make this choice for each of the n elements, by the Product Rule there are 2^n total subsets of $\{1, 2, \ldots, n\}$. $\square$

Example 7. How many subsets S of $\{1, 2, \ldots, n\}$ are there such that $|S|$ is odd?

Solution. For each element i ($1 \leq i \leq n-1$) we have two choices: Either i is in S or it is not in S. At this point, we consider $|S|$. If $|S|$ is odd, we must not include n in S. On the other hand, if $|S|$ is even (so far), we have to include n in S to satisfy the condition that $|S|$ is odd. In either case, we have only one choice for what to do with n. By the Product Rule, this implies there are $2^{n-1} \cdot 1 = 2^{n-1}$ subsets S of $\{1, 2, \ldots, n\}$ are there such that $|S|$ is odd. $\square$

Notice that this solution does not work when $n = 0$; certainly there are not 2^{-1} subsets of $\{\}$ with an odd number of elements. It is good to get in the habit of watching out for cases like this. If you are writing a solution on an exam, make sure you say that you are assuming $n > 0$.

Example 8. A dessert chef prepares the dessert for every day of a week starting with Sunday. The dessert each day is either cake, pie, ice cream, or pudding. The same dessert may not be served two days in a row. There must be cake on Friday because of a birthday. How many different dessert menus for the week are possible?

(2012 AMC 12B)

Solution. We start with Friday, since we know cake must be served that day. This implies that on Saturday, the dessert served cannot be cake, so we have 3 choices for that day's dessert. Similarly when we work backwards from Friday to Thursday, we see we have 3 choices for the dessert on Thursday (anything but cake). Then Wednesday we may select any of the 3 desserts not served Thursday, and so on back to Sunday. Since we have 3 choices for each day (aside from Friday), by the Product Rule we have $3^6 = 729$ possible menus. □

Note that there is something slightly subtle about our use of the Product Rule in Example 7. The Product Rule only requires that at each step in our chain of choices that we have the same *number* of possible choices at that point in our decision chain. What those specific options are does not matter. In Example 7, though the *set* of desserts allowed might change based on particular choices we make, for each day (besides Friday) the *number* of possible desserts is always exactly 3.

There are other ways to solve this problem as well. For example, we could have started with Monday and worked forward. Although this can work, it is much harder and involves some casework. (Try it if you don't believe us.) There are often several correct ways to solve counting problems, and it is always a good idea to consider different possible approaches.

Example 9. A large cube is painted green and then chopped up into 64 smaller congruent cubes. How many of the smaller cubes have at least one face painted green?

(Alabama ARML team selection)

Solution. We use complementary counting and determine how many cubes have no green faces. To have no green faces, a small cube must have been on the interior of the large cube. The large cube is $4 \times 4 \times 4$ with respect to the small cubes, so the interior of this cube is a $2 \times 2 \times 2$ group of small cubes. This is $2^3 = 8$ small cubes with no green faces, so there are $64 - 8 = 56$ small cubes with at least one face painted green. □

Example 10. Suppose $n \geq 2$ is a positive integer with prime factorization $n = p_1^{\alpha_1} p_2^{\alpha_2} \cdots p_k^{\alpha_k}$ where the p_i are prime numbers and the α_i are positive integers. How many factors does n have?

Solution. Recall that a number x is a factor of n if n is divisible by x. In order for this to be the case, the prime factorization of x must be $x = p_1^{\beta_1} p_2^{\beta_2} \cdots p_k^{\beta_k}$ where $0 \leq \beta_i \leq \alpha_i$ for each i. This means we have $\alpha_1 + 1$ choices for the value of β_1, $\alpha_2 + 1$ choices for the value of β_2, and so on. Applying the Product Rule, this tells us that the total number of divisors of n is $(\alpha_1+1)(\alpha_2+1)\cdots(\alpha_k+1)$. As an example, consider $20 = 2^2 \cdot 5^1$.

By our logic, 20 should have $(2+1)(1+1) = 6$ factors. They are $1, 2, 4, 5, 10,$ and 20. □

Example 11. Let n and k be positive integers. Count the number of k-tuples $(S_1, S_2, \ldots, S_k)$ of subsets of S_i of $\{1, 2, \ldots, n\}$ subject to each of the following conditions separately (i.e., the three parts are independent problems).

(a) The S_i's are pairwise disjoint.

(b) $S_1 \cap S_2 \cap \cdots \cap S_k = \emptyset$.

(c) $S_1 \cup S_2 \cup \cdots \cup S_k = \{1, \ldots, n\}$.

Solution.

(a) Consider a particular element $j \in \{1, 2, \ldots, n\}$. In order for the S_i's to be pairwise disjoint, j can be in at most one of $S_1, \ldots, S_k$. This is a total of $k+1$ possibilities (one for each subset and one for the possibility of j being in none of the subsets) for each of the n elements, so by the Product Rule there are $(k+1)^n$ k-tuples such that the S_i's are pairwise disjoint.

(b) Again consider a particular element $j \in \{1, 2, \ldots, n\}$. For each of the S_i we have 2 options: either j is in S_i or it is not. Thus there are a total of 2^k possible combinations of the subsets S_i that j could appear in. There is only 1 case that would violate our condition; the case where j is contained in every S_i. Thus there are $2^k - 1$ valid placements for each of the n elements of $\{1, \ldots, n\}$. Thus by the Product Rule there are $(2^k - 1)^n$ k-tuples satisfying $S_1 \cap S_2 \cap \cdots \cap S_k = \emptyset$.

(c) This is actually very similar to the previous part! There is only 1 case that would violate our condition; the case where j is contained in none of the S_i. Thus there are $2^k - 1$ valid placements for each of the n elements of $\{1, \ldots, n\}$. Thus by the Product Rule there are $(2^k - 1)^n$ k-tuples satisfying $S_1 \cup S_2 \cup \cdots \cup S_k = \{1, \ldots, n\}$. □

Chapter 2

Permutations and Combinations

Many counting problems involve arranging objects in certain places. Though the general idea is simple, these problems can vary greatly depending on the rules for what types of arrangements are allowed. In this chapter and the next we will look in depth at some general subtypes of these problems. We will see various things filling the roles of "objects" and "places." Throughout our discussion, the only rule that remains constant is that our "places" are distinguishable.

Distinguishable or Indistinguishable? Though our places will always be distinguishable, our objects may or may not be. For example, suppose we have two balls and two bins. How many ways are there to place our balls in our bins?

The answer here depends on whether or not our balls are distinguishable. If our balls are identical (indistinguishable), there are three ways to place them in the bins that look different to us: both balls in the left bin, both balls in the right bin, or one ball in each bin. However, if one of our balls is blue and the other is red, there are four possibilities for how we could see the balls distributed: we still have the case where both balls are in the left bin and both balls are in the right bin, but now we also have the case where the red ball is in the left bin and the blue ball is in the right bin and the case where the blue ball is in the left bin and the red ball is in the right bin.

Is repetition allowed? Another possibility we must consider is whether we can assign multiple objects to the same place. Above, we assumed that our bins were large enough to contain multiple balls each. Suppose we add the restriction that each bin must either be empty or contain exactly one ball. In this case, there is only one way to place two indistinguishable balls in our bins (one in each) and two ways to assign two distinguishable balls (red in the left bin and blue in the right or vice versa).

How can we tell which case applies? This is where counting problems can get tricky. Through the examples in these chapters and your own practice, you will learn to recognize hints that problem-writers give and become more comfortable identifying your objects and places (or balls and bins). If you are unable to figure out which case to use, all you can do is clearly state your assumptions and solve the problem as best you can.

We will now look at problems of four basic types—three in this chapter and one in the next. Our cases end up giving us four important basic forms that will show up again and again. We can find these forms using the Product Rule and some careful thinking to ensure we don't overcount. You will use these formulae often enough that you will likely end up memorizing them. However, knowing how to derive these basic forms will help you remember them better, allow you to more effectively determine which to apply to a particular situation, and get you used to the style of thinking that will allow you to solve more complicated counting problems.

To help demonstrate the forms in question, suppose there is a new ice cream shop in town that offers n different flavors of ice cream. To start attracting customers, they decide to run a series of special promotions.

- **Opening Day: distinguishable scoops, flavor repetition allowed**

    On opening day, the shop decides to offer k scoops of ice cream on a cone for a discounted price. Since the scoops are on a cone, they have an order, and thus are distinguishable. For example, having a scoop of vanilla on top of a scoop of chocolate is different from chocolate on top of vanilla. There are no restrictions, however, on what flavors you can get; you could pick k different flavors or k of the same flavor or anywhere in between. How many different cones are possible with these criteria?

Solution. Let's just start with a cone and build up our scoops! How many choices do we have for the flavor of the first scoop? There are n flavors, so we have n choices. What about the next scoop? Since there are no restrictions on flavor choice, we still have n choices. In fact, for every scoop we add, we have exactly n possibilities for what flavor it will be. This means in total, by the Product Rule there are

$$\underbrace{n \cdot n \cdots \cdots n \cdot n}_{k \text{ factors}} = n^k$$

possible cones we could make. □

The same method applies to a version of the classic "balls and bins" problem we alluded to before:

Permutations and Combinations

Example 12. How many ways are there to place k distinguishable balls into n distinguishable bins? (Distinguishable might mean that the balls are numbered from 1 to k and the bins are numbered from 1 to n.)

Solution. For our first ball, there are n choices for which bin to assign it to. Similarly, there are n choices for the second ball, and so on for every one of our k balls. By the Product Rule, this gives us a total of n^k ways to place balls in bins. □

We can relate this exercise to our ice cream count by imagining the balls are scoops of ice cream and the bins represent different flavors. When our objects are distinguishable and more than one object can appear in a particular place, the answer is always n^k.

- **Second Day: distinguishable scoops, no flavor repetition allowed**

Going back to our ice cream shop, suppose the owners realize that they ended up running out of the popular flavors way too early on opening day. To avoid this problem, the shop changes the conditions of the promotion for the next day. While customers can still get cones with k scoops, they can only have one scoop of any particular flavor.

Let's start with a special case. Suppose we're going to get a cone with n scoops (the same as the number of flavors). Since we can't repeat flavors, this means each flavor is used exactly once, and all we have to do is put them in order. How many ways are there to do this?

A definition will help us to get started.

Definition 2. A *permutation* of an ordered list of distinct objects is an ordered list with the same objects, but possibly in a different order.

For example, each of the following is a permutation of the ordered list $(1, 2, 3)$:

$$(1, 2, 3) \quad (1, 3, 2) \quad (2, 1, 3) \quad (2, 3, 1) \quad (3, 1, 2) \quad (3, 2, 1).$$

Solution. As before, we start from the bottom and build our cone up scoop by scoop. How many flavor choices do we have for the bottom-most scoop? We could pick any of the n available flavors, so there are n possibilities. What about the next scoop up? Now that we have our bottom scoop, we know that we can't pick that flavor, leaving us with only $n - 1$ flavors to choose from. Likewise, at the third scoop we have only $n - 2$ options since 2 flavors have been used up. This pattern continues till we reach the top scoop; since this is scoop n, we have used all but the last flavor and thus have only 1 option for which flavor to select. Applying the Product Rule, this gives us

$$n \cdot (n-1) \cdot (n-2) \cdot \ldots \cdot 3 \cdot 2 \cdot 1 \text{ possible cones.} \quad \square$$

The product above is so common in combinatorics that we give it its own name and notation. In particular, we call the product of the integers from 1 to n "n factorial" and denote it as $n!$. We can think of $n!$ as the number of ways to order n distinct objects. We say that $0! = 1$. This makes sense as there is only one way to arrange 0 objects (or create a 0 scoop ice cream cone), namely by doing nothing. (Note: you could also think of this in the sense that $0!$ is the "empty" product, and thus equals 1. Alternatively, note that $(n-1)! = n!/n$ giving $0! = 1!/1 = 1$.)

Now we focus on the general case where $k \leq n$. That is, how many ice cream cones can be made with k scoops from n total flavors, where no flavor may be used more than once?

Solution. Just as when we had exactly n scoops, there are n possibilities for the flavor of the bottom-most scoop. The next scoop up we have $n-1$ choices as we're not allowed to use the same flavor as the bottom scoop. This pattern continues. What, then, is the lowest number in our product? Since our first term is n, our second term is $n-1$, our third term is $n-2$ and so on. Thus our kth term will be $n-(k-1) = n-k+1$. This gives us, in general,

$$\underbrace{n \cdot (n-1) \cdot (n-2) \cdots (n-k+2) \cdot (n-k+1)}_{k \text{ factors}} = \frac{n!}{(n-k)!}$$

possible cones that suit our criteria. □

Some authors refer to this construction as a "falling power" or "falling factorial" and give it a notation such as $n^{\underline{k}}$ or $(n)_k$. We are not using these notations in this book, but it helps to remember that the expression $n!/(n-k)!$ is not something overly complicated—it is just a product of k factors, not much different from an ordinary power.

There is also an analogous balls and bins problem for permutations. We can see that $\frac{n!}{(n-k)!}$ counts the number of ways to place k distinguishable balls in n distinguishable bins if each bin can contain at most one ball.

- **Third Day: indistinguishable scoops, no flavor repetition allowed**

After two days of perilously stacking scoops on cones, the workers at the shop need a break. For Day 3, they opt to keep the flavor restriction in place (no flavor may be used more than once), but replace the cones with large bowls. These bowls are large enough that the order of the ice cream scoops does not matter, as a customer can use a spoon to eat whichever scoop he or she wants at any time (as opposed to a cone where the ice cream will topple over if eaten in the wrong order). A cone with a scoop of chocolate below scoop of vanilla would be considered different from a cone with a scoop of vanilla below a scoop of chocolate on top, but there is only one bowl counted containing one scoop of chocolate and one scoop of vanilla. Given these new rules, how many different k-scoop bowls are there?

Solution. Denote the answer to this question by $\binom{n}{k}$ (pronounced "n choose k"); that is, $\binom{n}{k}$ is the number of different bowls of ice cream we could make picking k scoops from n flavors, none of which may be used more than once. Our goal is to find a formula for $\binom{n}{k}$.

Let's think back to Day 2. We still had the rule of no more than one scoop of any flavor, but we cared about order then. The result was a total of $\frac{n!}{(n-k)!}$ different cones.

Suppose then that we are given some bowl of ice cream containing k scoops, each of which is a different flavor. How many different ways are there to place these scoops on a cone? This is simply a permutation of our k scoops, and we know from our special case above that there are $k!$ ways to arrange k distinct objects. We see that the total number of k-scoop cones of ice cream from n flavors where each flavor is used at most once is exactly the number of ways to select k flavors, then arrange those k flavors in some order on the cone. Thus, by the Product Rule, we see

$$\binom{n}{k} \cdot k! = \frac{n!}{(n-k)!}.$$

It follows that

$$\Rightarrow \binom{n}{k} = \frac{n!}{k!(n-k)!}$$

giving us our formula for $\binom{n}{k}$. $\binom{n}{k}$ is called a *binomial coefficient* (we will see why later). □

The value $\binom{n}{k}$ also counts the number of ways to place k indistinguishable balls into n distinguishable bins where each bin's capacity is at most one ball (again, imagine each bin corresponds to a particular flavor of ice cream and each ball represents a single scoop of ice cream). Another common application of $\binom{n}{k}$ is the following:

Example 13. Show that $\binom{n}{k}$ is the number of k element subsets of the set $\{1, \ldots, n\}$.

Solution. Imagine that we give each of our n flavors of ice cream a number. Then a particular k-scoop cup of ice cream where no flavor is repeated corresponds to exactly one k-element subset of $\{1, \ldots, n\}$.

See if you can come up with a direct argument to justify why this is true. Your argument will be very similar to the ice cream example. □

In general, we note that $\binom{n}{k}$ counts what we call combinations:

Definition 3. A *combination* is a subset of a group of distinct objects. (Order does not matter; $\{1, 2, 4\}$ is the same combination as $\{4, 2, 1\}$)

Now that we've established formulae for permutations and combinations, let's look at some examples where we can apply them.

Example 14. In how many ways can n people be seated in a circle? Two arrangements are considered the same if in both arrangements, each person's left neighbor is the same and each person's right neighbor is the same.

Solution. Since we consider two arrangements the same if we have rotational symmetry, we can use one person as a reference point and build our seating arrangement out from there. First we will choose who is sitting to the right of the reference person; there are $n-1$ choices for this seat since it cannot be our original reference person. Then we choose who is to the right of that person; we have $n-2$ choices since two people are already seated. We continue in this way around the table to get $(n-1) \cdot (n-2) \cdots 1$ so the answer is $(n-1)!$.

Notice that once we have our reference person, any seating corresponds to a unique permutation of the remaining $n-1$ people. We could also have used this observation to get the answer $(n-1)!$ directly. A third method would be to observe that there are n arrangements that are considered the same due to rotational symmetry. Thus we want to count one out of every n arrangements of n people, yielding $n!/n = (n-1)!$. □

We now define the notion of a lattice path. North-East lattice paths in particular prove to be very useful in combinatorics.

Definition 4. Suppose we are standing at the origin of a plane (i.e., at $(0,0)$). We proceed to move in the plane by taking horizontal or vertical unit steps. For example, after our first step we will end up at $(1,0), (0,1), (-1,0)$, or $(0,-1)$. A sequence of such steps is called a *lattice path*. If we restrict our possible directions to only right ("East") or up ("North") steps, we have a *North-East lattice path*.

We start here by finding a basic count for these paths.

Example 15. How many North-East lattice paths are there from $(0,0)$ to (a,b) where a and b are positive integers?

Solution. In order to reach (a, b), we must take exactly a steps to the right and b steps up (and thus $a+b$ total steps). We can think of this as choosing a numbers from the set $\{1, \ldots, a+b\}$, and for each i chosen, step i will be to the right. Once we have determined which steps are to the right, the remainder must be upward. We know the number of size a subsets from a set of size $a+b$ is $\binom{a+b}{a}$, giving us the number of lattice paths from $(0,0)$ to (a,b). □

Permutations and Combinations

The solution to Example 14 illustrates an important technique for counting problems which we also used informally in some earlier examples. We showed that to every North-East lattice path from $(0,0)$ to (a,b), we could associate a subset of $\{1, 2, \ldots, a+b\}$ with a elements, and conversely every such subset comes from a lattice path. This is an example of a *bijection*. A bijection is a one-to-one and onto matching between any two sets. More formally:

Definition 5. A bijection between two sets X and Y is a function $f : X \to Y$ such that for every $y \in Y$ there is a unique $x \in X$ with $f(x) = y$. In this case, we also say that f is a bijection from X to Y, and that X and Y are in bijection. Obviously, if two sets are in bijection, then they have the same number of elements. (This notion can even be used to talk about infinite sets having the same cardinality.)

We will often use bijections to relate the objects in a particular problem to some set that we have previously counted. This is a favored technique among mathematicians: taking a new problem and reducing it to one they already know how to solve!

Example 16. Let's play poker! In all of the following parts, we are talking about five card hands from a normal 52 card deck. "Suit" refers to whether a card is a spade, heart, diamond or club. "Rank" refers to the value of the card, ranging from ace, the numbers 2 through 10, jack, queen, and king. The order of cards in a poker hand does not matter.

(a) How many different poker hands are there?

(b) How many different hands contain the ace of spades?

(c) How many ways can you get a flush (i.e., all cards in your hand are the same suit)?

(d) How many ways can you get four of a kind (i.e., for a particular rank, you have that card in all four possible suits)?

(e) How many ways can you get a straight that is not a straight flush (a straight occurs when you have five consecutively ranked cards. There is no looping around - queen, king, ace, 2, 3 is not a straight - but an ace may count either as a "1" - below 2 - or as the highest rank - above king)?

(f) How many "junk" hands (i.e. hands with nothing better than a high card) are there in poker? Such a hand cannot be a flush, and all five cards must have different ranks (but not be a straight).

Solution.

(a) A poker hand is 5 cards chosen from the total 52 cards in the deck. Order of cards in a hand does not matter, so there are $\binom{52}{5}$ poker hands total.

(b) If we have the ace of spades in our hand, there are 4 other cards in our hand which must be chosen from the remaining 51 cards in the deck. Thus, there are $\binom{51}{4}$ hands containing the ace of spades.

(c) There are 4 ways to choose a suit for our flush. There are 13 cards in a given suit, so there are $\binom{13}{5}$ ways to pick the ranks of the five cards in our flush for a total of $4 \cdot \binom{13}{5}$ hands that are a flush.

(d) There are 13 possibilities for which rank we have four of. Since we must have all four suits of this rank, we just need to determine the final card in our hand. There are 48 other cards in the deck so there are $48 \cdot 13$ hands with four of a kind.

(e) We will count all straights, then subtract off the number of straight flushes. There are 10 possibilities for the lowest rank card in a straight (once we have chosen the lowest, the remaining ranks are determined). For each of the five cards in the hand, there are 4 choices for suit so we get $10 \cdot 4^5$ straights. However we need to subtract off straight flushes. There are still 10 possibilities for the lowest rank card in the straight, but now there are only 4 choices for suit since all cards in the hand must be of the same suit. Thus, the number of hands that are straights but not straight flushes is $10 \cdot 4^5 - 10 \cdot 4$.

(f) We can choose 5 different ranks in $\binom{13}{5}$ ways. However, we know there are 10 choices of ranks that will give us a straight. Thus there are $\binom{13}{5} - 10$ ways to select our ranks. Next, note that there are 4^5 ways to choose the suits of our 5 cards. There are 4 ways we could end up having all 5 cards the same suit, so there are overall $4^5 - 4$ possibilities for suits. Applying the Product Rule, we see that the number of junk hands is

$$\left(\binom{13}{5} - 10\right) \cdot (4^5 - 4). \qquad \square$$

Example 17. The figure below contains a total of 120 triangles of al sizes (that is, if we draw in all the lined and all n points along the horizontal base). What is the value of n?

(IdeaMath)

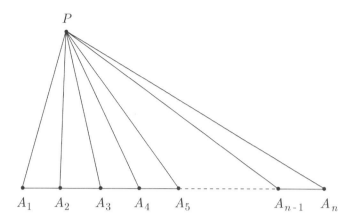

Solution. Note that a triangle in the figure is defined by three vertices, one of which is P, and the other two being different points on the horizontal base. Thus the number of triangles in the figure is exactly the number of pairs of vertices from the horizontal base. There are n such points; the number of ways to choose two of these (without worrying about order, since the order we choose the vertices does not change what triangle we get) is $\binom{n}{2}$. We also know there are 120 triangles so

$$\binom{n}{2} = \frac{n!}{2!(n-2)!} = \frac{n(n-1)}{2} = 120$$

which we can rearrange to obtain

$$n^2 - n - 240 = (n-16)(n+15) = 0.$$

Since n must be positive, we conclude that $n = 16$. □

Chapter 3

Stars and Bars and Multinomials

In Chapter 2 we covered three key counting situations. Here we will cover the case we missed, namely selecting a set of items (order does not matter) with repetition allowed. This case turns out to be trickier than the other three. In addition to this case, we also introduce multinomial coefficients in this chapter.

Let's go back to our ice cream metaphor.

- **Fourth Day: indistinguishable scoops, flavor repetition allowed**

On the final day of the promotion, the store opts to remove the flavor restriction and allow customers to select bowls with as many scoops of a given flavor as they want. There are still n flavors to choose from. How many k-scoop bowls meet these criteria?

Solution. It is tempting to try something similar to what we did in deriving our formula for $\binom{n}{k}$; in particular, we could divide our earlier formula for cones with repeated flavors by $k!$ to get $n^k/k!$. However, we can quicky see that this formula is incorrect; in fact, for some values of n and k it isn't even an integer! The problem here is that $k!$ is the number of ways to arrange k *distinct* objects. If our flavors are not distinct, there will not be $k!$ different cones using that particular flavor combination. In the extreme case when all the scoops are the same flavor, there is only one possible cone, so dividing this case by $k!$ makes no sense.

Instead we use a method known as "stars and bars" (also sometimes referred to as "balls and urns" or "sticks and stones"). Our goal is to make a bijection between the objects we are counting and arrangements of a given number of stars and bars. In the case of the ice cream shop, we do this as follows: We use k stars, each representing a single scoop of ice cream. We place bars as dividers between the stars to indicate the flavors of the scoops. For example, all stars appearing before the first bar would correspond to scoops of flavor 1, between the first and second bar would be scoops of flavor 2, and so on. This allows us to have multiple scoops of a single flavor (by having multiple stars between two bars) or no scoops of a flavor (when we have two bars next to each other).

As an example, suppose we have 3 flavors: coffee, mint chip, and chocolate. If we are interested in a bowl with 5 scoops of ice cream, we could denote the bowl with 2 scoops of coffee, 2 scoops of mint chip, and 1 scoop of chocolate by

$$\underbrace{\star\star}_{\text{coffee}} \mid \underbrace{\star\star}_{\text{mint chip}} \mid \underbrace{\star}_{\text{chocolate}}$$

Note that 2 bars are enough for 3 flavors. We can similarly translate any stars and bars diagram into a corresponding bowl of ice cream with some number of scoops. Consider the following two examples:

$$\star\star\star\star \mid \star\star \mid \star\star\star \quad \text{and} \quad \star \mid\mid \star\star \; .$$

Labeling these with the appropriate flavors gives us

$$\underbrace{\star\star\star\star}_{\text{coffee}} \mid \underbrace{\star\star}_{\text{mint chip}} \mid \underbrace{\star\star\star}_{\text{chocolate}} \quad \text{and} \quad \underbrace{\star}_{\text{coffee}} \mid \underbrace{}_{\text{mint chip}} \mid \underbrace{\star\star}_{\text{chocolate}} \; .$$

From this we can see that the first diagram represents 4 scoops of coffee, 2 scoops of mint chip, and 3 scoops of chocolate. The second diagram represents 1 scoop of coffee, 0 scoops of mint chip, and 2 scoops of chocolate.

Now let's generalize. To separate our diagram into n regions (flavors), we will need $n-1$ bars. For k scoops we need k stars. We can imagine having $n+k-1$ spaces to fill with our k stars and $n-1$ bars. We want to pick k slots where we will place stars. Note that choosing where the stars are completely determines the order, as every slot that does not have a star has a bar.

How many ways are there to choose the slots for our stars? We put exactly one object in each slot, so we cannot choose a single slot multiple times. Also, every slot we pick is filled with an identical star, so the order in which we choose the slots does not matter. Thus, we want to know the number of ways to choose k items from $n+k-1$ distinct objects without order or repetition, so we know that we have

$$\binom{n+k-1}{k}$$

possible stars and bars arrangements. Since each arrangement uniquely determines an ice cream bowl, this is exactly the number of possible k-scoop bowls from n total flavors where flavor repetition is allowed. □

Stars and Bars and Multinomials

Let's look at an explicit example here. If we have a bowl with 5 scoops and 3 allowed flavors, then our formula gives $\binom{7}{2} = 21$ possibilities. Is this right? Let's check. We are counting arrangements of 5 stars and 2 bars. The possibilities are as follows:

```
★★★★★||     ★★★★|★|     ★★★|★★|     ★★|★★★|     ★|★★★★|     |★★★★★|     ★★★★||★
★★★|★|★      ★★|★★|★     ★|★★★|★     |★★★★|★     ★★★||★★      ★★|★|★★     ★|★★|★★
|★★★|★★      ★★||★★★     ★|★|★★★     |★★|★★★     ★||★★★★      |★|★★★★     ||★★★★★
```

so we confirm there are indeed 21 of them.

Once again we note a balls-and-bins parallel for this scenario. We can use stars and bars to count the number of ways to distribute k indistinguishable balls among n distinguishable bins. Each ball goes in exactly one bin, but a bin may contain any number of balls including zero.

Multinomial Coefficients. One last (for now) important combinatorial concept to talk about! We've discussed how many ways there are to create ice cream cones from a selection of n flavors where we can repeat flavors however we wish. But what if we have an added restriction: that someone has told us how many scoops of each flavor we must have? In how many ways can we arrange these scoops on a cone?

Example 18. Suppose we have 3 flavors (vanilla, chocolate, and strawberry). We are told we must use v scoops of vanilla, c scoops of chocolate, and s scoops of strawberry. We'll call the total number of scoops k (so $v + c + s = k$). How many possible cones are there with these restrictions?

Solution. One way to think about this is to deal with the flavors one at a time. There are k scoops, of which v are vanilla. We can choose which v scoops will be vanilla in $\binom{k}{v}$ ways. We now have $k - v$ scoops left to be either chocolate or strawberry. Of these, we choose c scoops to be chocolate, which can be done in $\binom{k-v}{c}$ ways. This leaves us with $k - v - c = s$ scoops which will be strawberry. Taking the product of these gives us

$$\binom{k}{v} \cdot \binom{k-v}{c} = \frac{k!}{v!(k-v)!} \cdot \frac{(k-v)!}{c!s!} = \frac{k!}{v!c!s!}$$

Notice that the order in which we assigned flavors did not matter. That is, instead of choosing which scoops will be vanilla, then which scoops will be chocolate, and finally assigning the rest to be strawberry, we could have assigned strawberry then vanilla then chocolate. We would still get the same result. □

Though we used 3 flavors here, this argument can be extended to any number of flavors! In general, the number of ways to arrange k objects from a total of n different kinds of objects where there are k_1 of object type 1, k_2 of object type 2, ..., and k_n of object type n (where $k = k_1 + k_2 + \cdots + k_n$) is denoted

$$\binom{k}{k_1, k_2, \ldots, k_n} = \frac{k!}{k_1! k_2! \cdots k_n!}$$

We call this a *multinomial coefficient*.

When you have a multinomial coefficient with only two types of objects, you would get

$$\binom{n}{k, n-k} = \frac{n!}{k!(n-k)!} = \binom{n}{k}$$

so binomial coefficients are a special case of multinomial coefficients.

Now let's look back at our previous example and see if we can justify why this formula holds more directly (i.e., without using binomial coefficients). Both methods for deriving the formula are equally valid, and knowing both will help add to your mathematical toolbox for when you approach a new problem. The sort of argument we will see now is often helpful in adjusting to avoid overcounting.

Solution. [Alternate Solution for Example 17] Imagine we add food coloring to our ice cream so that we can differentiate between scoops of ice cream that are the same flavor. We will dye our scoops so that all our scoops of ice cream are "different" (i.e., every scoop of vanilla is dyed a different color, and a green scoop of vanilla is different from a blue scoop of vanilla). Then we have k distinct objects to order on our cone; we know from our previous discussions that the number of possible cones is $k!$.

However, in reality we aren't interested in crazy colored ice cream, we just care about the order of the flavors. This means we overcounted the possible cones, since having our blue vanilla scoop on the bottom with the green vanilla scoop on top is actually the same (taste-wise) as having the green scoop on bottom with the blue vanilla scoop on top. How many total possible color orderings are there for all of the vanilla scoops? Since we have v scoops of vanilla, there are $v!$ such orderings. Similarly there are $c!$ orderings of the colored chocolate scoops and $s!$ orderings of the colored strawberry. By the Product Rule, this means that for a specific flavor ordering, there are $v!c!s!$ colored orderings. Since we wish to count each flavor ordering exactly once, we need to divide by $v!c!s!$, yielding

$$\frac{k!}{v!c!s!} = \binom{k}{v, c, s}$$

as we found before. □

Stars and Bars and Multinomials

In the last couple sections we've covered a lot of important combinatorial formulae. Let's take a moment to summarize what we've seen so far in a table, focusing on the balls-and-bins example.

	No repetition	Repetition allowed
Distinguishable balls	$\dfrac{n!}{(n-k)!}$	n^k
Indistinguishable balls	$\binom{n}{k}$	$\binom{n+k-1}{k}$

By repetition, we refer to whether a bin is allowed to contain multiple balls.

Lastly, not represented in the table we have multinomial coefficients, which are used when we are given a fixed group of objects, some of which are identical, and must order these objects. Multinomial coefficients are also used when we are given a fixed set of labels, some of which are identical, and must assign these labels to a set of distinct objects.

Part of the challenge of solving a counting exercise if determining which of the above quantities (if any) are applicable to your problem. It can often be helpful to start by describing in English the process you will use to count what you are interested in. In the solutions to the example problems, this is generally our first step before we try to find mathematical expressions.

Once you have a plan, you will need to identify what is playing the roles of balls and bins in your problem, then work from there. For example, in our ice cream examples the scoops are the balls and the flavors are the bins. In the previous chapter, we considered an example with poker hands. In that case the balls would be the five slots in our hand and the bins would be a particular card in the deck.

We now explore some examples involving multinomial coefficients and stars and bars.

Example 19.

(a) How many "words" can be formed using all of the letters from the word AWESOMEMATH? (Note: In the context of a combinatorial problem, a word is just a collection of letters; it does not have to be an actual word you could find in the dictionary.)

(b) How many "words" can be formed using 5 of the letters from the word AWESOMEMATH?

Solution.

(a) One way of approaching this problem is to think of it in terms of multinomials. We have 11 total slots for our letters, and we need to fill these slots with a total of 2 A's, 2 E's, 2 M's, and 1 each of W, S, O, T, H. The number of ways to accomplish this will be

$$\binom{11}{2,2,2,1,1,1,1,1} = \frac{11!}{2!2!2!} = 4,989,600.$$

(b) We will look at three different cases here: the case where no letter is repeated, the case where one letter (either A, E, or M) is used twice, and the case where two letters (again from A, E, or M) are used twice. You can check that no other cases are possible.

- Case 1: Suppose we have no repeated letters. Then we are asking how many ways there are to order five out of eight total distinct objects, which we know to be $8!/3!$.

- Case 2: Suppose we have one pair of repeated letters. There are three ways to choose which letter is repeated from A, E, and M. There are then seven types of letters to choose our remaining three characters from; there are $\binom{7}{3}$ ways to do this. We then must arrange our five characters, one of which is a repeat. There are $\binom{5}{2,1,1,1}$ ways to do this, so overall there are $\binom{7}{3}\binom{5}{2,1,1,1}$ arrangements in this case.

- Case 3: Suppose we have two pairs of repeated letters. There are $\binom{3}{2}$ ways to choose which two letters from A, E, and M are repeated. We then must choose one of the remaining six character types to be our final letter. There are then $\binom{5}{2,2,1}$ ways to arrange our letters, so overall there are $\binom{3}{2} \cdot 6 \cdot \binom{5}{2,2,1}$ arrangements of this type.

Adding these all together, find

$$\frac{8!}{3!} + 3\binom{7}{3}\binom{5}{2,1,1,1} + 3 \cdot 6 \binom{5}{2,2,1} = 13560 \text{ words.} \qquad \square$$

Stars and Bars and Multinomials 25

Example 20. We arrange three blue plates, three red plates, and two green plates in a row.

(a) How many distinguishable arrangements are there?

(b) How many of the arrangements have red plates at both ends of the row?

(c) How many of the arrangements have all of the blue plates next to each other?

(d) How many of the arrangements have no two blue plates next to each other?

Solution.

(a) We use a multinomial coefficient. We have 8 total plates with distinguishable groups of size 3, 3, and 2 yielding

$$\binom{8}{3,3,2} = \frac{8!}{3!3!2!} = 560.$$

(b) We can satisfy this condition by forcing two red plates to be on either end of the row and creating an arrangement of the remaining plates between them. We are then essentially asking for the number of distinguishable ways to arrange three blue plates, one red plate, and two green plates in a row. Again using the multinomial coefficients, we find the number of arrangements to be

$$\binom{6}{3,2,1} = \frac{6!}{3!2!1!} = 60.$$

(c) To satisfy this condition, we can treat all three of these blue plates as a single super-plate unit. Under this logic, we are essentially arranging 1 blue super-plate, 3 red plates, and 2 green plates. Once again using multinomial coefficients, we have the number of arrangements as

$$\binom{6}{3,2,1} = \frac{6!}{3!2!1!} = 60.$$

(d) We begin by arranging only the red and green plates. This can be done in $\binom{5}{3}$ ways (we simply choose which spots will contain red plates and place the green plates in the remaining two spots). We have now defined 6 spots between the red and green plates and on the ends of the row. Each of these spots may contain at most one blue plate in order to satisfy the given condition, so we will choose 3 of them to each contain a blue plate. This can be done in $\binom{6}{3}$ ways, giving us a total of

$$\binom{5}{3}\binom{6}{3} = 10 \cdot 20 = 200$$

arrangements meeting the given criteria. □

Though the first three parts of the problem could be solved using multinomial coefficients, the constraints in part (d) are such that we cannot simply jump right in with our usual multinomial approach. Instead, we must pause and consider what *is* applicable in these circumstances.

Though we can attempt to make general statements about when certain formulae are applicable, this is an example of why we still must think very carefully and critically before applying the forms we have learned. It also gives a good example of how describing the procedure for building an arrangement of the desired type can help in figuring out the count. ("First, we arrange the red and green plates. Then, we place the blue plates among them so that they are not next to each other." This was probably not the first approach that occurred to you! This shows the importance of considering different approaches.)

Example 21. Find the number of ordered triples (a, b, c) of integers such that $0 \leq a \leq b \leq c \leq 11$.

Solution. Notice that given 4 (not necessarily distinct) integers between 0 and 11 inclusive, there is exactly one assignment of these numbers to a, b, c that is valid under our constraints (i.e., the assignment that satisfies $0 \leq a \leq b \leq c \leq 11$). As such, our goal is to determine the number of multisets of three integers selected from 0 to 11 inclusive with repetition allowed (A *multiset* is simply a collection of elements where some elements may appear multiple times. As with sets, order does not matter in a multiset). We can think of this as a stars and bars problem. Our stars will represent the three numbers we are choosing. We will have 11 bars, dividing the space into 12 regions, one for each integer from 0 up to 11. This is 14 total objects, so the number of arrangements (and thus the number of ordered triples) is $\binom{14}{3} = 364$. □

Example 22. How many nonnegative integral solutions does the equation

$$w + x + y + z = 11$$

have?

Solution. We can think of this as a stars and bars problem where each star represents a "unit" and the bars divide the stars up into the variables w, x, y, z.

In particular, the value of w will be the number of stars before the first bar, the value of x will be the number of stars between the first and second bars, the value of y will be the number of stars between the second and third bars, and the value of z will be the number of stars after the third bar. Note that this forces each variable to be a nonnegative integer. Since we need 11 total units, we have 11 stars and since we have 4 variables we have 3 bars. Overall there are $\binom{14}{3} = 364$ ways to arrange 11 stars and 3 bars, so this is the number of nonnegative integral solutions to our equation. □

But wait... our solution to Example 21 is exactly the same as our solution to Example 20. It turns out these two examples are really the same problem! We view a, b, and c as partial sums; that is, a is x, b is $x+y$, and c is $x+y+z$. w does not come into play, but it is fully defined by x, y, and z anyways.

It is very useful to look for connections between new problems and problems we already know how to solve. In the following examples, we will use bijections to relate what we want to count to problems of the same general form as Example 21.

Example 23. How many positive integral solutions does the equation

$$w + x + y + z = 15$$

have?

Solution. We will use a clever change of variables to relate to our previous example. In particular, let $W = w - 1, X = x - 1, Y = y - 1$, and $Z = z - 1$. We choose this assignment since w, x, y, z are positive integers exactly when W, X, Y, Z are nonnegative integers. Rearranging and substituting into our equation, we have

$$w + x + y + z = W + 1 + X + 1 + Y + 1 + Z + 1 = W + X + Y + Z + 4 = 15.$$

Subtracting from both sides we have

$$W + X + Y + Z = 11$$

where W, X, Y, Z are nonnegative integers. Then from what we learned in Example 21, we know there are $\binom{14}{3} = 364$ solutions. □

Without directly using our result from Example 21, here is an alternative way to solve Example 22:

Solution. [Alternate Solution for Example 22] We go back to stars and bars. We have 15 stars, one for each unit. We need to divide these units among our four variables, so we need three bars. However, unlike before, we have restrictions on where our bars may appear. In particular, we cannot have a bar on either end (as this would imply w or z was 0). Additionally, no two bars can be next to each other. Then there are 14 possible places to insert bars, and we may place at most one bar in each spot. Thus our answer is again $\binom{14}{3} = 364$. □

Example 24. The expression $(x+y+z)^{10}$ is expanded and simplified. How many terms are in the resulting expansion?

Solution. A term in the expansion $(x+y+z)^{10}$ is of the form $kx^a y^b z^c$ where k is some constant, and $a+b+c = 10$ (with a, b, c all nonnegative integers). We know from Example 21 that this is a stars and bars problem with 10 stars and 2 bars, so there are $\binom{12}{2}$ terms. □

Example 25. Eli, Joy, Paul, and Sam want to form a company; the company will have 16 shares to split among the 4 people. The following constraints are imposed:

- Every person must get a positive integer number of shares, and all 16 shares must be given out.

- No one person can have more shares than the other three people combined.

Assuming that shares are indistinguishable, but people are distinguishable, in how many ways can the shares be given out?

(2014 HMMT)

Solution. We use a similar approach to our alternate solutions from Example 22 along with complementary counting. First, we write 16 stars to represent the 16 shares. We insert 3 bars to partition the shares into Eli's part, Joy's part, Paul's part, and Sam's part. As in Example 22, we can't put bars on the end or multiple bars next to each other so that every person gets at least one share. This leaves us to choose 3 from 15 available places to insert the bars, which can be done in $\binom{15}{3} = 455$ ways.

In how many of the arrangements does Eli get 9 or more shares? This happens if there are 8 or more stars before the first bar. This leaves 7 possible slots for our bars, so there are $\binom{7}{3}$ ways this could happen. The same result occurs for Joy, Paul, and Sam, so the answer is $\binom{15}{3} - 4\binom{7}{3} = 315$ ways to split up the shares. □

Example 26. We have 20 marbles, each of which is either yellow, blue, green, or red. Assuming marbles of the same color are indistinguishable, at most how many marbles can be blue, if the number of ways in which we can arrange the marbles into a straight line is 1140?

Solution. Let the number of yellow, blue, green, and red marbles be y, b, g, and r respectively. We know $y + b + g + r = 20$. We also know that the number of ways to arrange the marbles in a line is 1140. Since we have several objects of 4 types that we need to arrange, this is a multinomial problem. We know there are

$$\binom{20}{y,b,g,r} = \frac{20!}{y!b!g!r!} = 1140 = 2^2 \cdot 3 \cdot 5 \cdot 19$$

ways to arrange our marbles. Note that since 19 is a prime factor of 1140, we know none of $y!, b!, g!,$ or $r!$ can contain a factor of 19, since our numerator 20! has only one factor of 19 to begin with. This implies $b < 19$. Similarly, since 1140 has no factor of 17, one of $y!, b!, g!, r!$ has a factor of 17. Since we want b to be as large as possible, we can say $b \geq 17$. This leaves us with two possible values to check: $b = 18$ or $b = 17$.

If we let $b = 18$, we know that

$$\frac{20!}{y!b!g!r!} \leq \frac{20!}{18!} = 20 \cdot 19 = 380 < 1140$$

so $b < 18$. Thus, $b = 17$. To confirm this, we note that

$$\frac{20!}{17!3!0!0!} = 1140,$$

giving us a valid distribution of marbles (17 blue and 3 of some other color). Thus, the maximum number of marbles that could be blue is 17. □

Chapter 4

Principle of Inclusion-Exclusion

Sometimes, the set that we are trying to count can be expressed as the union of several other sets. If the sets are disjoint, we can apply the Sum Rule to determine our final count. But what if the sets overlap? This is where the Principle of Inclusion-Exclusion (PIE) comes into play.

Let's look at some smaller cases to motivate the general form of this principle. Suppose we have two sets A and B and want to find the size of $A \cup B$. When we add $|A|$ and $|B|$ we end up counting any element that is in both A and B (i.e., in $A \cap B$) twice! To avoid overcounting, we subtract off the size of that intersection to yield

$$|A \cup B| = |A| + |B| - |A \cap B|. \qquad \text{(PIE for 2 sets)}$$

That wasn't so bad. But what happens when we have three sets instead of two? Things get slightly more complicated. As an example, suppose we have a class of students. Each student in this class likes some combination of armadillos, budgerigars, and capybaras (and every student likes at least one of these animals). Let A be the set of students who like armadillos, B be the set of students who like budgies, and C be the set of students who like capybaras. We would like to know how many students there are in our class, that is we want to find $|A \cup B \cup C|$.

We start with $|A| + |B| + |C|$ since this ensures every student gets counted at least once. What gets overcounted when we compute this sum? We can use a Venn Diagram to help us figure this out. The number in each region tells us how many times students in that region have been counted.

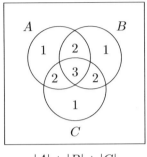
$$|A| + |B| + |C|$$

We can see there's definitely some overcounting going on. In particular, students who liked two types of animals have been counted twice, and students who like all three animals have been counting three times! To correct for this overcounting, let's try what we did with two sets and subtract off $|A \cap B|$. We will also subtract off $|A \cap C|$, then $|B \cap C|$. These are the Venn Diagrams representing the result after each of those steps:

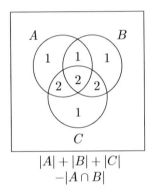
$$|A| + |B| + |C|$$
$$-|A \cap B|$$

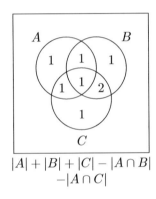
$$|A| + |B| + |C| - |A \cap B|$$
$$-|A \cap C|$$

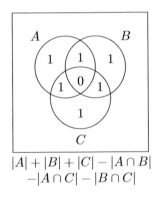
$$|A| + |B| + |C| - |A \cap B|$$
$$-|A \cap C| - |B \cap C|$$

We're close! Now we just have to make sure our centermost region (i.e., the set of students who like all three animals) is counted. To achieve this, we can add back in the size of the intersection of all three sets and everything will have been counted exactly once. To summarize, we've discovered that

$$|A \cup B \cup C| = |A| + |B| + |C| - |A \cap B| - |A \cap C| - |B \cap C| + |A \cap B \cap C|.$$
(PIE for 3 sets)

Huzzah! But what about even more sets? As you might suspect, there are even more rounds of adding and subtracting to correct for under and overcounting. Fortunately there is a pattern to how PIE works; we have seen the beginning of it in our examples with two and three sets. To get the size of the union of several sets, we...

- start with the sizes of the sets;
- subtract the sizes of the two-set intersections;

- add the sizes of the three-set intersections;
- subtract the sizes of the four-set intersections;
- and so on.

We subtract the sizes of all the even-set intersections and add the sizes of all the odd-set intersections until we get to the end (i.e. the intersection of all the sets). More formally:

Theorem 3. (Principle of Inclusion-Exclusion) *If we have n sets $A_1, A_2, \ldots, A_n$ then*

$$|A_1 \cup A_2 \cup \cdots \cup A_n| = \sum_{i=1}^{n} |A_i|$$
$$- \sum_{i<j} |A_i \cap A_j|$$
$$+ \sum_{i<j<k} |A_i \cap A_j \cap A_k|$$
$$- \cdots + (-1)^{n-1} |A_1 \cap \cdots \cap A_n|$$

Let's look at how we would apply the Principle of Inclusion-Exclusion to some specific problems. The last three example in this chapter deserve special attention. They are all important theorems in their own right—about counting "onto" functions, the totient function in number theory, and "derangements." Each is proven using PIE!

Example 27. At a particular school, there are 145 students who like pie (the dessert), 103 students who like π (the number), and 78 students who like PIE (the Principle of Inclusion-Exclusion). There are 54 students who like pie and π, 42 students who like π and PIE, and 33 students who like pie and PIE. There are 19 students who like all three. Assuming every student likes at least one of pie, *pi*, and PIE, how many students are there at the school?

Solution. This problem is a straightforward application of the Principle of Inclusion-Exclusion. Our sets will be people who like pie (A), people who like π (B), and people who like PIE (C). The Principle of Inclusion-Exclusion tells us

$$|A \cup B \cup C| = |A| + |B| + |C| - |A \cap B| - |A \cap C| - |B \cap C| + |A \cap B \cap C|.$$

Substituting the appropriate values, we have:

$$|A \cup B \cup C| = 145 + 103 + 78 - 54 - 33 - 42 + 19 = 216 \text{ students.} \qquad \square$$

Example 28. In a class there are 40 students. 14 of them like math, 16 like physics, 11 like chemistry. 7 students like both math and physics, 8 students like both physics and chemistry, 5 students like both math and chemistry, and 4 students like all the subjects. How many students do not like any of the subjects?

Solution. This problem uses the Principle of Inclusion-Exclusion along with complementary counting; these techniques are often used together. Let A be the set of students who like math, B be the set of students who like physics, and C be the set of students who like chemistry. Then the set of students who like none of these subjects is $40 - |A \cup B \cup C|$. We compute $|A \cup B \cup C|$ using PIE to get

$$40 - |A \cup B \cup C| = 40 - (14 + 16 + 11 - 7 - 8 - 5 + 4) = 15$$

so there are 15 students who do not like any of the subjects. □

Example 29. How many integers between 1 and 240 inclusive are multiples of 4 or 6?

Solution. We know there are $240/4 = 60$ multiples of 4 between 1 and 240 inclusive and $240/6 = 40$ multiples of 6 between 1 and 240 inclusive. If we add these two numbers, then we have counted multiples of $\text{lcm}(4,6) = 12$ twice, since they are counted in each category. To remedy this we subtract off the $240/12 = 20$ multiples of 24 to get $60 + 40 - 20 = 80$ integers. □

When we have n sets, there is one term for every non-empty intersection of a subset of the sets $A_1, A_2, \ldots, A_n$. This means there are $2^n - 1$ terms on the right side of the PIE formula (Theorem 3). If we are unlucky, this can make PIE impractical when n is large.

Luckily, sometimes we will have a lot of symmetry. In this case, large values of n should not scare us. For example, for any k, we know there are $\binom{n}{k}$ terms representing k-set intersections. What if these are all the same size? Then we can count any one of them and multiply by $\binom{n}{k}$, making the problem much easier. We see this put to use in the next example.

Example 30. The four sets A, B, C, and D each have 400 elements. The intersection of any two of the sets has 115 elements. The intersection of any three of the sets has 53 elements. The intersection of all four sets has 28 elements. How many elements are there in the union of the four sets?

(2007 Purple Comet)

Solution. This is another straightforward application of PIE. We know each set has 400 elements and there are $n = 4$ sets. Each intersection of two sets

Principle of Inclusion-Exclusion 35

has 115 elements and there are $\binom{4}{2}$ such pairs. Each intersection of three sets has 53 elements and there are $\binom{4}{3}$ such triples. Finally, the intersection of all four sets has 28 elements. Overall, this gives us

$$|A \cup B \cup C \cup D| = 4 \cdot 400 - \binom{4}{2} \cdot 115 + \binom{4}{3} \cdot 53 - 28 = 1094.$$

(You might wonder whether $\binom{4}{1}$ and $\binom{4}{4}$ should appear in this formula. Where would they go?) □

Example 31. In how many ways can we arrange the letters in the word MEMENTO if no two of the same letter may appear consecutively in an arrangement?

Solution. We know using multinomial coefficients that the total number of ways to arrange the letters of MEMENTO is $7!/(2!2!)$. The number of ways for the two M's to appear consecutively is $6!/2!$ since we can just treat the two M's as a single unit and move them together. By similar logic, there are $6!/2!$ ways for the two E's to appear together. When we subtract these two numbers, we find we have overcorrected for the case where both M's are together *and* both E's appear together so we must add that back. Treating the M's as a single group and the E's as a single group leaves us with 5 objects and thus $5!$ arrangements. Overall, this gives us

$$\frac{7!}{2!2!} - 2\frac{6!}{2!} + 5! = 660$$

valid arrangements. □

Example 32. How many integer solutions are there to the system

$$w + x + y + z = 11 \text{ such that } 0 \le w, x, y, z \le 4?$$

Solution. We have seen from Example 21 that there are $\binom{14}{3} = 364$ nonnegative integer solutions to $w + x + y + z = 11$. This takes care of the restriction that $0 \le w, x, y, z$. We will use complementary counting to determine how many nonnegative integer solutions to our equation are such that at least one of w, x, y, z is greater than 4.

Suppose at least one of our variables is too large. There are 4 ways to choose which of our variables is violating the upper bound of 4. We can assign 5 units to the chosen variable, leaving 6 additional units to be distributed. There are $\binom{9}{3}$ ways to distribute these units among all 4 variables.

Now suppose at least two of our variables are too large; there are $\binom{4}{2}$ ways to choose which. When we assign 5 units to each of these variables, we are left with 1 more unit. There are $\binom{4}{3}$ ways to distribute this. We notice that it is impossible for more than two variables to violate the upper bound while all of our variables are nonnegative, since this would give us a sum of at least 15. Thus, applying the Principle of Inclusion-Exclusion, we have

$$\binom{14}{3} - 4 \cdot \binom{9}{3} + \binom{4}{2}\binom{4}{3} = 52$$

solutions such that $0 \le w, x, y, z \le 4$. (Does it surprise you that this excludes most of the original 364 solutions?) □

Problems can be more difficult when they contain variables rather than actual numbers, but they are still approachable as long as we keep the general pattern of the Principle of Inclusion-Exclusion in mind. Of course, if you find the variables muddling up your counting, you can always try substituting some small values for the variables to give yourself a sense of what is going on before moving to the general case.

Example 33. Recall that a function f from X to Y is said to be *onto* if for every $y \in Y$, there exists some $x \in X$ such that $y = f(x)$ (i.e., every element of Y is mapped to by f applied to some element of X). Prove that the number $s_{m,n}$ of onto functions $f : X \to Y$, where $|X| = n$ and $|Y| = m$ (both positive) is given by the expression:

$$s_{m,n} = m^n - \binom{m}{1}(m-1)^n + \binom{m}{2}(m-2)^n - \cdots + (-1)^{m-1}\binom{m}{m-1}1^n.$$

(A common synonym for "onto" is "surjective".)

Solution. We have two variables in this problem, which may at first seem intimidating. Let's try plugging in a small value for m (say $m = 4$) and see what happens.

It's not immediately clear how we could go about counting onto functions. However, we know that a function that is *not* onto must "miss" mapping to some element of Y. Counting these functions seems more approachable, so we try complementary counting and the Principle of Inclusion-Exclusion. Let A_i be the set of all functions that miss the ith element of Y for $1 \le i \le 4$. Then what we want to count is

total # of functions $- |A_1 \cup A_2 \cup A_3 \cup A_4|$.

First, how many functions are there from X to Y where $|Y| = 4$? For each $x \in X$ we must choose some element of Y to be $f(x)$, so there are 4 possible assignments of $f(x)$ for each x. The Product Rule tells us we have 4^n functions from X to Y.

Now we count the functions that are not onto. This implies at least one element of Y is missed. Suppose a is some element of Y and we wish to count all functions such that a is not mapped to by anything in X. This means we only have 3 possible values for each $f(x)$, so we end up getting 3^n functions that miss a. How many possibilities are there for what a could be? There are $\binom{4}{1} = 4$ possible values of a. Thus overall we have counted $\binom{4}{1} \cdot 3^n$ functions that miss at least one element of Y.

We know that we have overcounted several functions here (namely those that miss more than one element), so we follow the Principle of Inclusion-Exclusion and continue our count. Next we count functions that miss two elements of Y (say a and b). There are 2^n functions that miss both a and b, and there are $\binom{4}{2}$ possible choices for what a and b are, giving us a total of $\binom{4}{2} \cdot 2^n$ functions. Similarly, we find there are $\binom{4}{3} \cdot 1^n$ functions missing three of our elements. There are no functions missing all four elements; objects in X have to map to something!

Applying our usual Principle of Inclusion-Exclusion formula yields

$$s_{4,n} = 4^n - \binom{4}{1} \cdot 3^n + \binom{4}{2} \cdot 2^n - \binom{4}{3} \cdot 1^n,$$

which agrees with the formula given in the problem statement when $m = 4$. Now let's see if we can generalize this.

As with our $m = 4$ case, we will use complementary counting and the Principle of Inclusion-Exclusion to determine the number of onto functions. Since $|Y| = m$ and $|X| = n$, there are m choices for each x, and thus m^n total functions $f : X \to Y$.

We could go through the cases of at least 1 element being missed, at least 2 elements being missed, and so on as before. Let's see if we can speed this up a bit by considering the case of k elements being missed where k is some integer $1 \leq k < m$.

Suppose we have a function that misses at least some set S of k elements of Y. That is, $S \subset Y$, $|S| = k$, and for each $s \in S$, there is no $x \in X$ such that $f(x) = s$. This implies every element of X must be mapped by f to an element in $Y \setminus S$. There are $m - k$ such elements; thus there are $(m-k)^n$ functions that avoid S. Overall there are $\binom{m}{k}$ subsets S of size k. Applying the Principle of Inclusion-Exclusion, we obtain

$$s_{m,n} = \sum_{k=0}^{m} (-1)^k \binom{m}{k} (m-k)^n$$

$$= m^n - \binom{m}{1}(m-1)^n + \binom{m}{2}(m-2)^n - \cdots + (-1)^{m-1}\binom{m}{m-1} 1^n.$$

This is exactly what we were asked to prove. □

Example 34. Euler's Totient function, $\varphi(n)$ is the number of positive integers less than or equal to n that are relatively prime to n (i.e., that share no common factors with n greater than 1). If the prime factorization of n is $n = p_1^{\alpha_1} p_2^{\alpha_2} \cdots p_k^{\alpha_k}$ where the p_i are distinct primes and the α_i are positive integers, show that

$$\varphi(n) = n\left(1 - \frac{1}{p_1}\right)\left(1 - \frac{1}{p_2}\right) \cdots \left(1 - \frac{1}{p_k}\right).$$

Solution. This is a particularly challenging exercise, largely due to the number of variables floating around. Here especially it is useful to plug in actual numerical values of n to try to get a sense of the pattern. Before jumping to the general case, let us examine what happens when $n = 360$. We know the prime factorization of 360 is $2^3 \cdot 3^2 \cdot 5$. We will use complementary counting and the Principle of Inclusion-Exclusion to approach this problem.

Let us count how many positive integers less than or equal to 360 share a common factor greater than 1 with 360. Such numbers must be multiples of 2, 3, or 5, so we define A to be the set of positive multiples of 2 up to 360, B to be the set of positive multiples of 3 up to 360, and C to be the set of positive multiples of 5 up to 360. Our answer is then

$$\varphi(360) = 360 - |A \cup B \cup C|$$
$$= 360 - (|A| + |B| + |C| - |A \cap B| - |A \cap C| - |B \cap C| + |A \cap B \cap C|).$$

The set $A \cap B$ contains exactly those numbers that are multiples of 2 and 3, or multiples of $2 \cdot 3 = 6$. Similarly, $A \cap C$ contains the multiples of 10, $B \cap C$ contains the multiples of 15, and $A \cap B \cap C$ contains the multiples of 30. Plugging in the appropriate values, we see

$$\varphi(360) = 360 - \frac{360}{2} - \frac{360}{3} - \frac{360}{5} + \frac{360}{6} + \frac{360}{10} + \frac{360}{15} - \frac{360}{30}$$
$$= 360\left(1 - \frac{1}{2} - \frac{1}{3} - \frac{1}{5} + \frac{1}{2 \cdot 3} + \frac{1}{2 \cdot 5} + \frac{1}{3 \cdot 5} - \frac{1}{2 \cdot 3 \cdot 5}\right)$$
$$= 360\left(1 - \frac{1}{2}\right)\left(1 - \frac{1}{3}\right)\left(1 - \frac{1}{5}\right)$$

as expected. Now let us move to the case for general n.

As before, we count how many natural numbers less than or equal to n share a common factor greater than 1 with n. Any multiple of p_i will share a factor of p_i with n. There are $\frac{n}{p_i}$ natural numbers less than or equal to n that are multiples of p_i. Summing up over all values of i yields $\sum_{i=1}^{k} \frac{n}{p_i}$.

We know, however, that this is overcounting. By the Principle of Inclusion-Exclusion, we subtract off all numbers that are multiples of two of the p_i's. There are $\sum_{1 \leq i_1 < i_2 \leq k} \frac{n}{p_{i_1} p_{i_2}}$ such numbers. Then we add back those that are multiples of three of the p_i's to obtain $\sum_{1 \leq i_1 < i_2 < i_3 \leq k} \frac{n}{p_{i_1} p_{i_2} p_{i_3}}$.

We repeat this process till we arrive at multiples of all k primes in the prime factorization of n, yielding $(-1)^{k-1} \frac{n}{p_1 p_2 \cdots p_k}$ added to our sum. Subtracting our result from n, we find

$$\varphi(n) = n - \sum_{i=1}^{k} \frac{n}{p_i} + \sum_{1 \leq i < j \leq k} \frac{n}{p_i p_j} - \cdots + (-1)^k \frac{n}{p_1 p_2 \cdots p_k}.$$

Finally, this can be factored to yield

$$\varphi(n) = n \left(1 - \frac{1}{p_1}\right)\left(1 - \frac{1}{p_2}\right) \cdots \left(1 - \frac{1}{p_k}\right)$$

as desired. □

Example 35. A *derangement* of a set of objects is a permutation of those objects such that no object ends up in its orignal spot. For example, $4, 3, 1, 2$ is a derangement of $1, 2, 3, 4$.

Questions involving derangements are often posed as a story problem. For example, suppose that n homework assignments are randomly returned to n students. In how many ways could the papers be returned such that no student gets his or her own homework back (i.e., how many derangements of $1, \ldots, n$ are there)?

Solution. Unsurprisingly, we use the Principle of Inclusion-Exclusion and complementary counting! Let A_i be the set of arrangements where student i gets his or her own homework back (for $1 \leq i \leq n$). We want the total number of ways to return homework, minus $|A_1 \cup \cdots \cup A_n|$. We know that the number of ways to return the homework with no restrictions is simply the number of ways to permute n objects; that is, $n!$. Now let's determine the number of arrangements where at least one student gets his or her homework back.

Here is where we start using PIE. First we have to add the size of each individual set. Notice that since all the students are treated equally, we can take advantage of symmetry. For any one student i, $|A_i|$ (the number of ways he can get his own homework back) is $(n-1)!$ (since the remaining $n-1$ papers may be distributed in any way). Since there are n such sets, we add $n \cdot (n-1)!$.

Now what about the intersection of two sets $|A_i \cap A_j|$ (i.e., when students i and j get their own papers back)? We can distribute the remaining papers in any way, so there are $(n-2)!$ possibilities. How many intersections of two sets do we have? This is exactly the number of ways to pick a pair of students from our total of n students, which we know is $\binom{n}{2}$. Thus we subtract off $\binom{n}{2}(n-2)!$ from our running count.

As we continue working, this pattern continues. If k students get their own homework back, there are $\binom{n}{k}$ ways to choose which students those are and then $(n-k)!$ ways to distribute the rest of the homework. Ultimately, this ends up giving us:

$$n \cdot (n-1)! - \binom{n}{2} \cdot (n-2)! + \binom{n}{3} \cdot (n-3)! + \cdots + (-1)^{n-1}\binom{n}{n} \cdot 0!$$

Remember though that this is the complement of what we want to count, so we need to subtract it from the total possible ways to return papers (i.e., $n!$):

$$n! - n \cdot (n-1)! + \binom{n}{2} \cdot (n-2)! - \binom{n}{3} \cdot (n-3)! + \cdots + (-1)^n \binom{n}{n} \cdot 0!$$

$$= \sum_{k=0}^{n} (-1)^k \binom{n}{k} (n-k)!$$

We can simplify this algebraically to find the number of derangements of n objects is

$$\sum_{k=0}^{n} (-1)^k \binom{n}{k}(n-k)! = \sum_{k=0}^{n} (-1)^k \frac{n!}{k!(n-k)!}(n-k)!$$

$$= n! \sum_{k=0}^{n} \frac{(-1)^k}{k!}.$$

(This is the famous "derangements formula." If you have studied power series, the sum may look familiar to you. If not, you have many exciting mathematical treats to look forward to!) □

Chapter 5

Pascal's Triangle and the Binomial Theorem

Pascal's Triangle is a fascinating construction. In this chapter we will barely begin to describe its many wonders. Before this chapter is over, we will see several example of a powerful prood technique, and we will learn why the entries in Pascal's Triangle are called "binomial coefficients."

$$
\begin{array}{lccccccccc}
n=0: & & & & & 1 & & & & \\
n=1: & & & & 1 & & 1 & & & \\
n=2: & & & 1 & & 2 & & 1 & & \\
n=3: & & 1 & & 3 & & 3 & & 1 & \\
n=4: & 1 & & 4 & & \boxed{6} & & 4 & & 1
\end{array}
$$

Pascal's Triangle. The rows start with $n = 0$, and the entries in each row start with $k = 0$. For example, entry 2 of row 4 (the boxed entry) is $\binom{4}{2} = 6$.

The top row (row $n = 0$) has just one entry, 1. After that, each row n has $n+1$ entries, and the first and last entry (entries $k = 0$ and $k = n$) are each 1. Aside from these 1's, every entry is the sum of the two entries above it.

We see that entry k in row n is exactly $\binom{n}{k}$. (This is why we number our rows starting with 0 and the entries of each row starting with 0.)

Why is this the case? It turns out that when $n \geq 2$ and $0 < k \leq n$,

$$\binom{n}{k} = \binom{n-1}{k} + \binom{n-1}{k-1}. \qquad \text{(Pascal's Identity)}$$

Since binomial coefficients follow the same rule as we have used to construct the triangle (and since $\binom{n}{0} = \binom{n}{n} = 1$ for every n), every new entry we create is the appropriate binomial coefficient!

Of course, we still need to prove Pascal's Identity. There are many ways to prove it. One way is to use algebra, starting with the formula we saw earlier

$$\binom{n}{k} = \frac{n!}{k!(n-k)!}.$$

You can check this proof for yourself. Instead, we offer a combinatorial argument.

The idea behind the proof is to define a set, then count its members in two different ways. The two expressions we find must be equal since they count the same set! This is sometimes called the "double counting" method, and it is an essential technique in combinatorics.

Let's state our theorem formally.

Theorem 4. (Pascal's Identity) $\binom{n}{k} = \binom{n-1}{k} + \binom{n-1}{k-1}$ *for all* $0 < k \leq n$.

Some people define the expression $\binom{n}{k}$ even when k is not in the range from 0 to n by saying $\binom{n}{k} = 0$ whenever $k < 0$ or $k > n$. If we use this definition, Theorem 4 applies more generally: It is true when $n \geq 1$ and k is any integer. For simplicity (and so our counting makes sense), our proof will assume $0 < k \leq n$.

Proof. How many ways are there to form a committee of size k from a class of n students?

Answer 1: We need to pick k people from a set of n different people. Each person can be chosen at most once, and the order in which we choose the representatives does not matter. This gives us a result of $\binom{n}{k}$.

Answer 2: Let's focus on one particular student in the class (say Jenny). This is where our assumption that $n \geq 1$ is important, as otherwise we would not be able to select Jenny! There are two possibilities - either Jenny is on the committee, or she is not. If she is, we need to pick $k-1$ other students from the remaining $n-1$ students to complete the committee; we can do this in $\binom{n-1}{k-1}$ ways. Otherwise, she is not on the committee, and we must pick all k members from the remaining $n-1$ students, which can be done in $\binom{n-1}{k}$ ways. By the Sum Rule, this means the total number of ways to create a committee is $\binom{n-1}{k} + \binom{n-1}{k-1}$.

Since we have two expressions that count the same thing, they must be equal. Thus, we have shown that

$$\binom{n}{k} = \binom{n-1}{k} + \binom{n-1}{k-1}$$

as desired. $\square$

Committee-forming arguments like this one are a good technique to keep in mind when trying to prove combinatorial identities involving binomial coefficients. Another common type of argument involves North-East lattice paths, which we saw previously in Example 14. We can give an alternate proof of Pascal's Identity using a lattice path argument as follows:

Proof. How many North-East lattice paths are there from the origin $(0,0)$ to $(k, n-k)$?
Answer 1: We take a total of n steps in our path, of which k are to the right and $n-k$ are up. There are $\binom{n}{k}$ ways to choose which k of our steps are to the right; the other $n-k$ steps are up.
Answer 2: We look at two cases: the case where the last step is up and the case where the last step is to the right.

If the last step is up, the point we visit immediately prior to $(k, n-k)$ is $(k, n-k-1)$. The number of lattice paths from $(0,0)$ to $(k, n-k-1)$ is $\binom{n-1}{k}$ (we take a total of $n-1$ steps and choose k to be to the right).

If the last step is to the right, the point we visit immediately prior to $(k, n-k)$ is $(k-1, n-k)$. The number of lattice paths from $(0,0)$ to $(k-1, n-k)$ is $\binom{n-1}{k-1}$ (we take a total of $n-1$ steps and choose $k-1$ to be to the right). Since these are our only two cases, we apply the Sum Rule to find that there are $\binom{n-1}{k} + \binom{n-1}{k-1}$ total North-East lattice paths from $(0,0)$ to $(k, n-k)$.

Since we have two expressions that count the same thing, they must be equal. Thus, we have shown that

$$\binom{n}{k} = \binom{n-1}{k} + \binom{n-1}{k-1}$$

as desired. □

Pascal's Triangle has many interesting properties beyond Pascal's Identity. For example, let's look at what happens if we take the sum of a row of Pascal's Triangle. (We have seen part of this argument before in Example 5.)

Example 36. Use a counting argument to show that the sum of the elements of row n of Pascal's triangle is 2^n. That is,

$$\sum_{k=0}^{n} \binom{n}{k} = \binom{n}{0} + \binom{n}{1} + \cdots + \binom{n}{n} = 2^n.$$

Solution. How many committees (of any size) can be formed from a class of n students?
Answer 1: We know that the number of ways to form a committee of size k from a group of n people is $\binom{n}{k}$. Now we simply have to sum over all possible

values of k. Our committee could have as few as zero members or as many as all n people in the class, so k ranges from 0 to n. Thus we have

$$\binom{n}{0} + \binom{n}{1} + \binom{n}{2} + \cdots + \binom{n}{n} = \sum_{k=0}^{n} \binom{n}{k}.$$

<u>Answer 2:</u> For each person, we have two possibilities – either the student is on the committee or he is not. We must make this decision for each of the n students. By the Product Rule the number of ways we can do this is 2^n.

Since we have two expressions that count the same thing, they must be equal. Thus, we have shown that

$$\sum_{k=0}^{n} \binom{n}{k} = 2^n$$

as desired. □

Notice that we could just as easily have framed our proof here in terms of the question "How many subsets of $\{1, 2, \ldots, n\}$ are there?" (which is how we saw the quesiton posed in Example 5). How we describe the set of objects we are counting does not matter as long as it is the appropriate size. Generally you can come up with your set to count based on one side of the equation you are trying to prove, then the challenge is finding a counting procedure that will yield the other side.

Now let's turn our attention to the diagonals of Pascal's Triangle. We call this result the "Hockey Stick Theorem" (try circling all the numbers involved in the identity in Pascal's Triangle and you'll see why). There is an algebraic proof that can be done by appealing to Pascal's identity which we do not include here. We will, however, provide both a committee-forming proof as well as a lattice path version of the proof.

Theorem 5. (Hockey Stick Theorem) *If $n \geq k \geq 0$, then*

$$\binom{k}{k} + \binom{k+1}{k} + \binom{k+2}{k} + \cdots + \binom{n}{k} = \binom{n+1}{k+1}.$$

Proof. How many ways are there to pick a $k+1$ person hockey team from $n+1$ players (wearing jerseys numbered 1 to $n+1$)?
<u>Answer 1:</u> This is simply the number of ways to choose $k+1$ people from a set of $n+1$ distinct players, which we know is $\binom{n+1}{k+1}$.
<u>Answer 2:</u> Let's suppose the highest jersey number of a player on the team is $r+1$. Then we know that the other k players must have jersey numbers r or lower, so we can choose them in $\binom{r}{k}$ ways. We know $r+1$ must be at least $k+1$ (or else we won't have enough players on our team) and it can be as

large as $n+1$. Thus r ranges from k to n. Summing over the possible values of r gives us
$$\sum_{r=k}^{n}\binom{r}{k}=\binom{k}{k}+\binom{k+1}{k}+\cdots+\binom{n}{k}.$$

Since we have two expressions that count the same thing, they must be equal. Thus, we have shown
$$\binom{k}{k}+\binom{k+1}{k}+\binom{k+2}{k}+\cdots+\binom{n}{k}=\binom{n+1}{k+1}$$
as desired. □

For our lattice path proof, we will rewrite the identity in a way that is more convenient for this technique. Let $m = n - k$. Then our identity, in terms of m and k, becomes
$$\binom{k}{k}+\binom{k+1}{k}+\binom{k+2}{k}+\cdots+\binom{m+k}{k}=\binom{m+k+1}{k+1}.$$

Proof. How many North-East lattice paths are there from $(0,0)$ to $(k+1,m)$?
<u>Answer 1:</u> Our path must contain $k+1$ steps to the right and m steps up. There are $\binom{m+k+1}{k+1}$ ways to choose which steps will be to the right (the rest will be up).
<u>Answer 2:</u> We will break into cases based on the smallest ℓ such that $(k+1, \ell)$ is on our path (i.e. the first time our path reaches the line $x = k+1$). Since we define ℓ to be the smallest such value, we know the step immediately prior to reaching $(k+1, \ell)$ was to the right, so we count paths from $(0,0)$ to (k, ℓ). There are $\binom{\ell+k}{k}$ such paths. Since we only allow steps up or to the right, once the path reaches $(k+1, \ell)$ all remaining steps must be up so there is only one way to complete the path to $(k+1, m)$. Now ℓ could range from 0 to m, so summing over all possible values of ℓ, we have
$$\sum_{\ell=0}^{m}\binom{k+\ell}{k}=\binom{k}{k}+\binom{k+1}{k}+\binom{k+2}{k}+\cdots+\binom{m+k}{k}.$$

Since we have two expressions that count the same thing, they must be equal. Thus, we have shown
$$\binom{k}{k}+\binom{k+1}{k}+\binom{k+2}{k}+\cdots+\binom{m+k}{k}=\binom{m+k+1}{k+1}$$
as desired. □

Binomials and Multinomials. We can also use our knowledge of binomial coefficients to help us when we're looking at polynomials.

Theorem 6. (Binomial Theorem) *For a nonnegative integer* $n \geq 1$

$$(x+y)^n = \binom{n}{0}x^0y^n + \binom{n}{1}x^1y^{n-1} + \cdots + \binom{n}{n}x^ny^0 = \sum_{k=0}^{n}\binom{n}{k}x^ky^{n-k}.$$

Now we see the reason for the term "binomial coefficients"! They are the coefficients in the binomial expansion. We will see a proof by induction of this theorem later (see Example 61), but here we give a justification for why this is true based on counting principles.

Let's first examine a small case with $n=4$. We can think of $(x+y)^4$ as

$$(x+y)(x+y)(x+y)(x+y).$$

When we apply the distributive law to expand this out but do not simplify or gather like terms, we end up with something like

$x \cdot x \cdot x \cdot x + x \cdot x \cdot x \cdot y + x \cdot x \cdot y \cdot x + x \cdot x \cdot y \cdot y + x \cdot y \cdot x \cdot x + x \cdot y \cdot x \cdot y + x \cdot y \cdot y \cdot x + x \cdot y \cdot y \cdot y$

$+ y \cdot x \cdot x \cdot x + y \cdot x \cdot x \cdot y + y \cdot x \cdot y \cdot x + y \cdot x \cdot y \cdot y + y \cdot y \cdot x \cdot x + y \cdot y \cdot x \cdot y + y \cdot y \cdot y \cdot x + y \cdot y \cdot y \cdot y.$

Which of these terms will simplify to x^2y^2? Exactly those terms which contain two x's and two y's in some order. How many such orders are there? We have 4 total characters, and we wish to select 2 of them to be x's; we know we can do this in $\binom{4}{2}$ ways. This implies there are $\binom{4}{2}$ terms in our sum with two x's and two y's, and thus the coefficient on x^2y^2 will be $\binom{4}{2}$. Try checking the other coefficients for this case to see that they work out as well.

Now we generalize by consider $(x+y)^n$ where n is a nonnegative integer. What is the coefficient on x^ky^{n-k} in the expansion of this expression?

As before, we can imagine having n different "units" of $(x+y)$. When we expand this out, each term ends up being some number of x's and some number of y's (n characters in total) in some order. The terms that simplify to x^ky^{n-k} are exactly those consisting of k x's and $n-k$ y's in some order. The number of possible orders of these characters is the number of ways to choose k of the n characters to be x's, which we know is $\binom{n}{k}$. Thus, we end up seeing the product x^ky^{n-k} in our expansion $\binom{n}{k}$ times, so its coefficient is $\binom{n}{k}$.

There is a similar theorem for multinomials, called (intuitively enough), the *Multinomial Theorem*. It tells us how to expand out a multinomial to the nth power.

$$(x_1 + x_2 + \cdots + x_m)^n = \sum_{k_1+k_2+\cdots+k_m=n} \binom{n}{k_1, k_2, \ldots, k_m} x_1^{k_1} x_2^{k_2} \cdots x_m^{k_m}$$

The justification is similar to the argument for the Binomial Theorem, so we will not discuss it here. Instead, let's jump into some examples showing how we can apply the Binomial and Multinomial Theorems to problems we may encounter (and not just problems with polynomials!).

Example 37. Here we use the Binomial and Multinomial Theorems to find coefficients for terms in polynomials.

(a) What is the coefficient on the x^2y^3 term in the expansion of $(x+2y)^5$?

(b) What is the coefficient on the x^3y^3 term in the expansion of $(2x-3y)^6$?

(c) What is the coefficient on the x^3y^3 term in the expansion of $(x-2y+3)^8$?

Solution.

(a) By the Binomial Theorem, we know that the x^2y^3 term will be $\binom{5}{2}x^2(2y)^3$ so the coefficient will be $2^3 \cdot \binom{5}{2} = 80$, and the term is $80x^2y^3$.

(b) By the Binomial Theorem, we know that the x^3y^3 term will be $\binom{6}{3}(2x)^3(-3y)^3$ so the coefficient will be $2^3 \cdot (-3)^3 \cdot \binom{6}{3} = -4320$, and the term is $-4320x^3y^3$.

(c) By the Multinomial Theorem, we know that the x^3y^3 term will be $\binom{8}{3,3,2}x^3(-2y)^3 3^2$ so the coefficient will be $(-2)^3 \cdot 3^2 \cdot \binom{8}{3,3,2} = -40320$, and the term is $-40320x^3y^3$. $\square$

Example 38. Earlier we used a counting argument to prove

$$\sum_{k=0}^{n} \binom{n}{k} = \binom{n}{0} + \binom{n}{1} + \cdots + \binom{n}{n} = 2^n.$$

Now prove this identity using the binomial theorem.

Solution. The binomial theorem tells us that

$$(x+y)^n = \binom{n}{0}x^0y^n + \binom{n}{1}x^1y^{n-1} + \cdots + \binom{n}{n}x^ny^0.$$

Setting $x = y = 1$, this yields

$$2^n = (1+1)^n = \binom{n}{0}1^0 1^n + \binom{n}{1}1^1 1^{n-1} + \cdots + \binom{n}{n}1^n 1^0$$
$$= \binom{n}{0} + \binom{n}{1} + \cdots + \binom{n}{n}$$

as desired. $\square$

Example 39. What is the hundreds digit of 2011^{2011}?

(2011 AMC 10B)

Solution. This problem can be solved using a clever application of the multinomial theorem. Note that we can write 2011^{2011} as $(2000+10+1)^{2011}$. Then by multinomial theorem, we have

$$\sum_{a+b+c=2011} \binom{2011}{a,b,c} 2000^a 10^b 1^c.$$

Note that since we are interested in the hundreds digit, we can ignore any term with a positive power of 2000. This leaves us to deal with the case where $a = 0$. Rewriting this in terms of one variable (as we are used to with the binomial theorem) we have

$$\sum_{k=0}^{2011} \binom{2011}{k} 10^k 1^{n-k}.$$

Here again there are many terms we can ignore; any term with $k \geq 3$ will have a factor of $10^3 = 1000$ and thus will not contribute to the hundreds digit. This leaves us with

$$\binom{2011}{0} 10^0 + \binom{2011}{1} 10^1 + \binom{2011}{2} 10^2 = 1 + 20110 + 202105500$$

so the hundreds digit of 2011^{2011} will be 6. $\square$

In addition to the interesting properties within Pascal's Triangle and in the binomial expansion, binomial coefficients pop up in other contexts, as shown by the next example.

Example 40. A triangular array of squares has one square in the first row, two in the second, and in general, k squares in the kth row for $1 \leq k \leq 11$. With the exception of the bottom row, each square rests on two squares in the row immediately below (illustrated in the given diagram). In each square of the 11th row, a 0 or a 1 is placed. Numbers are then placed into the other squares, with the entry for each square being the sum of the entries in the two squares below it.

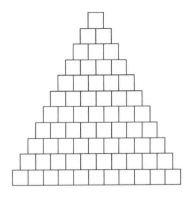

For how many initial distributions of 0's and 1's in the bottom row is the number in the top square a multiple of 3?

(2007 AIME II)

Solution. Label the elements in the bottom squares $a, b, c, d, e, f, g, h, i, j, k$ from left to right where each entry is either a 0 or 1. If we look at the leftmost element in the bottom row, we see it is a. The leftmost element in the next row up is $a + b$, then $a + 2b + c$ in the row above that, $a + 3b + 3c + d$ in the row above that, and $a + 4b + 6c + 4d + e$ in the next row up.

We notice that the coefficients on the variables in these entries correspond to elements of Pascal's triangle. In particular, the coefficients for the ℓth row correspond to the $(11 - \ell)$th row of Pascal's triangle for $1 \leq \ell \leq 11$. Thus the coefficients in the top square will match the 10th row of Pascal's triangle. This tells us the entry in this square will be

$$a + 10b + 45c + 120d + 210e + 252f + 210g + 120h + 45i + 10j + k.$$

Since we are only concerned with this number being a multiple of 3, let's pull out as many multiples of 3 as we can:

$$3(3b + 15c + 40d + 70e + 84f + 70g + 40h + 15i + 3j) + (a + b + j + k).$$

Since the first part will always be a multiple of 3, we can simply focus on $a + b + j + k$. Since each of these is 1 or 0, their sum must be either 0 or 3 if we want a multiple of 3. The sum is 0 only when a, b, j, k all equal 0. The sum is 3 when three of the four are 1 and the last is 0. This can happen in $\binom{4}{3} = 4$ ways (we choose which 3 will be 1). The remaining 7 variables can be assigned either 0 or 1 freely; this can be done in 2^7 ways. Thus our total distributions are $5 \cdot 2^7 = 640$.

We could compute the coefficients in many ways: by writing out the first 11 rows of Pascal's triangle, by using our formula $\binom{n}{k} = \frac{n!}{k!(n-k)!}$, etc. A neat shortcut here, since we are only interested in what their remainder is upon division by 3, is to write Pascal's triangle only keeping track of the remainder upon division by 3. For example, the first several rows would be

$$
\begin{array}{cccccccccc}
n = 0: & & & & & 1 & & & & \\
n = 1: & & & & 1 & & 1 & & & \\
n = 2: & & & 1 & & 2 & & 1 & & \\
n = 3: & & 1 & & 0 & & 0 & & 1 & \\
n = 4: & 1 & & 1 & & 0 & & 1 & & 1 \\
\end{array}
$$

This method for computing the remainder of the coefficients is appealing because it is much faster and less prone to arithmetical error. □

Chapter 6

Counting in More Than One Way

The double counting principle is one the most fundamental and important techniques in combinatorics. The general situation to picture a representation of this principle, is if the information given to you in the problem is stored in a table, then the sum of the information can be obtained in two ways, either adding up the columns or adding up the rows.

A more concrete way to visualize the principle is that we describe some set S and then count the elements in it in two ways; thus it reduce to a equation of the type **L.H.S = R.H.S**. One way would give the **L.H.S** and the other would give the **R.H.S**.

We use this technique earlier in chapter 5 to prove theorems 4 and 5 and to solve Example 35. This is one of the fundamental techniques and proofs using it we can generally call "combinatorial proofs". Combinatorial proofs give extra insight, are fun to write and to read. The only hard thing in these type of problems is to find the correct set S to count.

Let's start working through the examples and will a better sense of what this principle implies.

Example 41. Prove that
$$\tau(1) + \tau(2) + \ldots + \tau(n) = \left\lfloor \frac{n}{1} \right\rfloor + \left\lfloor \frac{n}{2} \right\rfloor + \ldots + \left\lfloor \frac{n}{n} \right\rfloor,$$
where $\tau(k)$ denotes the number of divisors of the positive integer k.

Solution. Let S be the set of ordered pairs (d, k) where $1 \leq d \leq k \leq n$ and d is a divisor of k.

On the one hand let's ask for a given k how many such d's there are? The answer is obvious; by definition it is $\tau(k)$. Thus summing up over all k's
$$|S| = \tau(1) + \tau(2) + \ldots + \tau(n).$$

On the other hand, switching perspective, if we fix d, how many such k's can we find with $d|k$ and $1 \leq k \leq n$? The answer is $\left\lfloor \frac{n}{d} \right\rfloor$. Thus again summing up over all choices of d,

$$|S| = \left\lfloor \frac{n}{1} \right\rfloor + \left\lfloor \frac{n}{2} \right\rfloor + \ldots + \left\lfloor \frac{n}{n} \right\rfloor.$$ □

Example 42. Prove that
$$\sum_{d|n} \varphi(d) = n,$$
where $\varphi(x)$ denotes Euler's totient function, namely it is equal to the number of positive integers coprime to x which are less than or equal to x.

Solution. We count the number of fractions in the set $A = \left\{ \frac{a}{n} \mid 1 \leq a \leq n \right\}$ in two ways. First it is obvious that there are n such fractions.

On the other hand, some of the fractions can be reduced and after this we realize that each of them has a representative of the type $\frac{p}{d}$ where $d|n$ and $(p, d) = 1$. Thus for each $d|n$ we have $\varphi(d)$ fractions with denominator d.

It follows that there are $\sum_{d|n} \varphi(d)$ fractions in A, so $\sum_{d|n} \varphi(d) = n$. □

Example 43. Let $a_1 \leq a_2 \leq \ldots \leq a_n = m$ be positive integers. Denote by b_k the number of those a_i for which $a_i \geq k$. Prove that

$$a_1 + a_2 + \ldots + a_n = b_1 + b_2 + \ldots + b_m.$$

Solution. Let us change the perspective on b_k. If we take the following n intervals $[1, \ldots a_1], (a_i, a_{i+1}]$ for $1 \leq i \leq n-1$, then every number k in $\{1, 2, \ldots, n\}$ lands in exactly one these intervals. For all j in the interval $[1, \ldots a_1]$, we have that $b_j = n$. Similarly, for each j in the interval $(a_i, a_{i+1}]$ (for $1 \leq i \leq n-1$), we have $b_j = n - i$. Thus

$$b_1 + b_2 + \ldots + b_m = na_1 + \sum_{i=1}^{n-1}(n-i)(a_{i+1} - a_i) = a_1 + a_2 + \ldots + a_n.$$ □

Example 44. Let $p_n(k)$ be the number of permutations of the set $\{1, 2, 3, \ldots, n\}$ which have exactly k fixed points. Prove that $\sum_{k=0}^{n} k p_n(k) = n!$.

(IMO 1987)

Solution. Look firs at the left hand side. This expression counts the number of fixed points over all permutations of the set. More precisely, it counts the number of pairs (π, x) where π is a permutation of $\{1, 2, \ldots, n\}$ and x is a fixed point of π. For each $k = 0, \ldots, n$ there are $p_n(k)$ permutations π with k fixed points, and each correspond to k pairs (π, x). So the whole number is given by the *L.H.S*.

Now if we switch perspective, we can count this same number of pairs by splitting according to x. For each x, to form a permutation that fixes x, the other numbers can be permuted in an arbitrary way. Thus there are $(n-1)!$ permutations that fix x. So in all there are $n \cdot (n-1)! = n!$ pairs.

The identity follows. □

Example 45. 200 students participated in a mathematical contest. They had 6 problems to solve. It is known that each problem was correctly solved by at least 120 participants. Prove that there must be 2 participants such that every problem was solved by at least one of these 2 participants.

(IMC)

Solution. We count in two ways the pairs of participants with the property that both of them did not solve one of the problems in the mathematical contest.

By hypothesis we know that for any problem there are at most 80 participants who did not solve it. Thus we get for each problem that there are at most $\binom{80}{2}$ pairs of participants who did not solve it. Thus there at most $6 \cdot \binom{80}{2}$ pairs. On the other hand since there are $\binom{200}{2}$ pairs of participants, and $6 \cdot \binom{80}{2} < \binom{200}{2}$, there must be at least a pair with the property for any of the 6 problems, one of them solved it. □

Example 46. Given p and q positive integers, prove that

$$\left\lfloor \frac{p}{q} \right\rfloor + \left\lfloor \frac{2p}{q} \right\rfloor + \ldots + \left\lfloor \frac{(q-1)p}{q} \right\rfloor = \left\lfloor \frac{q}{p} \right\rfloor + \left\lfloor \frac{2q}{p} \right\rfloor + \ldots + \left\lfloor \frac{(p-1)q}{p} \right\rfloor$$

where by $\lfloor x \rfloor$ denotes the integral part of the real number x.

Solution. This is a very typical problem for counting lattice point (points with both coordinates integers). We consider the cartesian coordinate plane and we are going to count the number of lattice points inside or on the boundary the triangle with coordinate $(0,0)$, $(p,0)$ and (p,q).

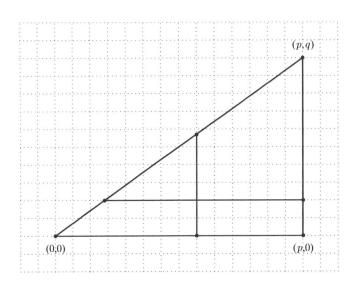

The two ways we are proceeding to count is doing it by horizontal lines and by vertical lines.

Thus take a point $(m, 0)$ with $1 \leq m \leq p-1$. If we draw a vertical line through this point, it meets the line from $(0,0)$ to (p,q) at the point with y-coordinate equal to $\frac{qm}{p}$. There are precisely $\left\lfloor \frac{qm}{p} \right\rfloor$ lattice points. In total we get $\sum_{m=1}^{p-1} \left\lfloor \frac{qm}{p} \right\rfloor$ lattice points.

Now for the horizontal lines, we take again a point of coordinates $(p, q-n)$ and draw the horizontal line inside the triangle until it meets the line from $(0,0)$ to (p,q). The meeting point has coordinates $\left(p - \frac{np}{q}, q-n\right)$. We thus get on each horizontal line of this type $\left\lfloor \frac{pn}{q} \right\rfloor$ lattice points. The total is now $\sum_{n=1}^{q-1} \left\lfloor \frac{pn}{q} \right\rfloor$ lattice points. □

Example 47. In a competition, there are a contestants and b judges, where $b \geq 3$ is an odd integer. Each judge rates each contestant as either "pass" or "fail". Suppose k is a number such that, for any two judges, their ratings coincide for at most k contestants. Prove that

$$\frac{k}{a} \geq \frac{b-1}{2b}.$$

(IMO 1998)

Solution. Let us draw a table in which we have b columns corresponding to the judges and a rows corresponding to the contestants. We also record in the cell (i,j) for $1 \leq i \leq a$ and $1 \leq j \leq b$, a p if the judge j passes the contestant i and f if he fails him.

We will count the pairs of agreements between votes, call it A, namely those of the type (p,p) and (f,f), in two ways, namely by rows and by columns.

By columns, we looking at votes cast by different judges. Using the hypothesis we know that for each pair out of $\binom{b}{2}$ of judges, their votes agree for at most k contestants so $A \leq k\binom{b}{2}$.

By rows, let us note that the the situations are independent for the contestant so we can count for one contestant and then multiply by a, the number of contestant. So let us consider the first contestant and say he obtained m passes and n fails. Note that $m + n = b$.

Then the number of agreements is

$$\binom{m}{2} + \binom{n}{2} = \frac{m^2 + n^2 - (m+n)}{2}$$

$$= \frac{(m+n)^2 + (m-n)^2 - 2(m+n)}{4} = \frac{b^2 - 2b + (m-n)^2}{4}$$

Since b is odd and m and n are integers it follows that $|m - n| \geq 1$ so the number of agreements for the first contestant is at least $\frac{(b-1)^2}{4}$.

Thus we get a lower bound on the number of agreements namely

$$A \geq a \cdot \frac{(b-1)^2}{4}.$$

Putting it all together we obtain that

$$a \cdot \frac{(b-1)^2}{4} \leq A \leq k\binom{b}{2} \text{ so } a \cdot \frac{(b-1)^2}{4} \leq k\binom{b}{2}$$

and we obtain the desired inequality, $\frac{k}{a} \geq \frac{b-1}{2b}$. □

Example 48. Let $A_1, A_2, \ldots, A_k$ be subsets of $\{1, 2, \ldots, n\}$ each of them with at least $\frac{n}{2}$ elements, and such that for $i \neq j$ we have $|A_i \cap A_j| \leq \frac{n}{4}$. Prove that

$$\left| \bigcup_{i=1}^{k} A_i \right| \geq \frac{k}{k+1} n$$

Solution. Let $\bigcup_{i=1}^{k} A_i = \{b_1, b_2, \ldots, b_m\}$. Our goal is to prove that $m \geq \dfrac{k}{k+1}n$.

We consider the following table with rows indexed by the sets $A_1, A_2, \ldots, A_k$ and columns indexed by the elements $b_1, b_2, \ldots, b_m$. We put a 1 in the cell (i, j) with $1 \leq i \leq k$ and $1 \leq j \leq m$ if b_j is an element of the set A_i. Let x_i the number of 1's in the column i, or equivalently it is the number of sets in which the element b_i appears with $1 \leq i \leq m$.

Now we have that $S = \sum_{i=1}^{m} x_i$ is equal to the sum of the numbers in the whole table, and doing the sum by rows we obtain that

$$\sum_{i=1}^{m} x_i = \sum_{i=1}^{k} |A_i| \geq \frac{nk}{2}.$$

Let us look further at $\sum_{1 \leq i < j \leq k} |A_i \cap A_j|$. This is equal to the number of pairs of 1's which are in the same column. But we can rewrite this in our notation as

$$\sum_{1 \leq i < j \leq k} |A_i \cap A_j| = \sum_{i=1}^{m} \binom{x_i}{2}.$$

By hypothesis we know that each intersection has at most $\dfrac{n}{4}$ elements so

$$\sum_{i=1}^{m} \binom{x_i}{2} \leq \frac{n}{4}\binom{k}{2}. \qquad (*)$$

Note that

$$\sum_{i=1}^{m} \binom{x_i}{2} = \frac{1}{2}\left(\sum_{i=1}^{m} x_i^2 - \sum_{i=1}^{m} x_i\right) \geq \frac{1}{2}\left(\frac{S^2}{m} - S\right) = \frac{S^2 - mS}{2m}$$

using Cauchy-Schwartz.

Since $S - \dfrac{m}{2} \geq \dfrac{nk}{2} - \dfrac{m}{2} \geq 0$, we have

$$S^2 - mS + \frac{m^2}{4} \geq \frac{n^2 k^2}{4} - \frac{mnk}{2} + \frac{m^2}{4}.$$

Hence

$$\sum_{i=1}^{m} \binom{x_i}{2} \geq \frac{n^2 k^2 - 2mnk}{8m}.$$

Combining this with (∗) we conclude that

$$\frac{n^2k^2 - 2mnk}{8m} \leq \frac{n}{4}\binom{k}{2}.$$

After a bit of manipulation this is equivalent to the given statement, namely

$$m \geq \frac{k}{k+1}n.$$

□

Finally we end with three examples which have a geometrical component also, which gives us some restrictions on our counts due to the euclidean geometry constraints which arise naturally.

Example 49. There are n points in the plane, such that no three of them are collinear. Prove that the number of triangles, whose vertices are chosen from these n points and whose area is 1, is not greater than $\frac{2}{3}(n^2 - n)$.

(Iran 2010)

Solution. Let's call the triangles with area 1 *good* and let T the number of such triangles. Let us look at the sides of such a good triangle. Each triangle has three sides, so the number of segments is equal to $3T$.

On the other hand if we have a given segment AB it is a part of at most four *good* triangles.

If we assume the contrary, then three of the points lie in same halfplane bounded by AB, say they are X, Y, Z. It is an elementary geometric fact that if $[ABC] = [ABD]$, where we use this notation for area, and C, D are on the same side of AB then $CD \| AB$. Thus $XY \| AB$ and $YZ \| AB$ so X, Y, Z are collinear and this contradicts the assumption of the problem.

We conclude that because of this the number of segments which appear in *good* triangles, is at most $4 \cdot \binom{n}{2}$, since this is the maximal number of segments we can draw between n points.

Thus $3T \leq 2n(n-1)$ so $T \leq \frac{2n(n-1)}{3}$.

□

Example 50. Let n, k be positive integers and let S be a set of n points in the plane such that no three are collinear and for every point $P \in S$ there are at least k points in S equidistant from P. Prove that $k < \frac{1}{2} + \sqrt{2n}$.

(IMO 1989)

Solution. Let us label the points $P_1, P_2, \ldots, P_n$. We consider the circles $\mathcal{C}_1, \mathcal{C}_2, \ldots, \mathcal{C}_n$ with centers P_i such that for each P_i a maximal set of points equidistant from P_i are on the circle $\mathcal{C}_i$.

We count triples (P_i, P_j, P_k) such that $j < k$ and P_j and P_k lie on $\mathcal{C}_i$. For each index i, there are at least k points on the circle $\mathcal{C}_i$. Hence there are at least $\binom{k}{2}$ triples beginning with P_i and at least $n\binom{k}{2}$ triples in total.

Now on the other hand, for each pair (P_j, P_k), if they both lie on a circle $\mathcal{C}_i$, then the center P_i of the circle must lie on the perpendicular bisector of $P_j P_k$. Since no 3 points of S are collinear, there are at most 2 triples ending in (P_j, P_k) and at most $2\binom{n}{2}$ triples in total.

Combining these two counts we have $n\binom{k}{2} \leq 2\binom{n}{2}$, or $k(k-1) \leq 2(n-1)$. Rewriting this we get $\left(k - \frac{1}{2}\right)^2 \leq 2n - \frac{7}{4} < 2n$, or $k < \frac{1}{2} + \sqrt{2n}$. □

Example 51. Let M be a set of $n \geq 4$ points in the plane, such that no three are collinear and not all of them lie on a circle. Find all functions $f : M \to \mathbb{R}$ with the property that for any circle C containing at least three points of S,

$$\sum_{P \in C \cap M} f(P) = 0$$

(Romania TST 1997)

Solution. Call the circles which contain at least three of the points from M *good*. Every triplet of the points belongs to exactly one *good* circle. We begin with a lemma

Lemma.
$$\sum_{P \in M} f(P) = 0.$$

Proof. For a triplet of points $T = \{P, Q, R\}$ let $f(T) = f(P) + f(Q) + f(R)$. Consider $\sum_{T \subset M} f(T)$, with T ranging over every possible triplet.

On one hand, every point of M is included in exactly $\binom{n-1}{2}$ triplets, so

$$\sum_{T \subset M} f(T) = \binom{n-1}{2} \sum_{P \in M} f(P).$$

On the other hand, suppose we choose a *good* circle $\mathcal{C}$ and count all the triplets which are entirely on it. Then if the circle has k points from M, each point will be included in $\binom{k-1}{2}$ triplets; thus

$$\sum_{T \subset M \cap \mathcal{C}} f(T) = \binom{k-1}{2} \sum_{P \in M \cap \mathcal{C}} f(P) = 0.$$

Now we simply sum over all *good* circles; since each triplet is included in exactly one *good* circle we have

$$\sum_{T\subset M} f(T) = 0$$

and combining these two equations proves the lemma. $\square$

Now let A and B be two arbitrary points and let S be the set of all *good* circles which pass through both of them. Not all of the points of M are concyclic, hence $|S| > 1$. Also, every point other than A and B which is in M is in exactly one circle of S, so

$$0 = \sum_{\mathcal{C}\in S}\sum_{P\in\mathcal{C}\cap M} f(P) = |S|(f(A)+f(B)) + \sum_{P\in M\setminus\{A,B\}} f(P).$$

Applying the lemma, we have $0 = (|S|-1)(f(A)+f(B))$; since $|S|>1$ we know $f(A)+f(B) = 0$.

The rest of the proof is easy: for three arbitrary points A, B, C we have

$$2f(A) = (f(A)+f(B)) + (f(A)+f(C)) - (f(B)+f(C)) = 0$$

so the only function that works is the zero function. $\square$

Chapter 7

Pigeonhole Principle

The *Pigeonhole Principle* states that if there are k holes and more than k pigeons, and every pigeon is in a hole, then there must be a hole that contains at least two pigeons. More generally, if there are k holes and n pigeons, some hole must contain at least $\lceil \frac{n}{k} \rceil$ pigeons and some other hole must contain at most $\lfloor \frac{n}{k} \rfloor$ pigeons.

Despite its simplicity, the Pigeonhole Principle can be applied in creative ways to prove some surprisingly complicated results. We approach a Pigeonhole Principle problem by identifying which objects represent our "pigeons" and which objects represent our "holes." Doing this generally makes up the bulk of the proof; once we have identified our objects, we apply the Pigeonhole Principle to reach our conclusion.

Example 52. Consider five points in a square with side length 2. Prove that no matter how these points are placed, some pair of them are no more than $\sqrt{2}$ apart.

When we begin thinking about this problem, we might consider the worst case scenario—a case in which the points are as far apart as possible. This seems to be when we place four points in the corners of the square and one in the center. Even in this case, the statement holds true. But this is far from a rigorous argument! How do we know we have found the worst case? Each case we consider is easy, but there are so many cases that it is difficult to cover them all.

The Pigeonhole Principle comes to the rescue.

Solution. The five points are the pigenos in this problem, but what are the holes? None are specifically outlined in the problem statement, so we must come up with our own. We partition the square into four unit squares by connecting the midpoints of opposite sides; each of these unit squares is a hole. By the Pigeonhole Principle, one of these four unit squares must contain at least two of the five points. In a unit square, the maximum distnce between two points is $\sqrt{2}$, so these two points are at most distance $\sqrt{2}$ apart as desired. □

(Even in this formal solution we have been a bit sloppy. We have not specified to which "hole" each of the boundary lines belongs. There are multiple fixes for this: Either we can define the "holes" more carefully so each boundary point is only included in one of the squares; or we can argue that it doesn't matter, since the Pigeonhole Principle still applies if objects can be placed in multiple holes.)

Example 53. Suppose there are n people at a conference and some of these people shake hands with one another. Show that there are at least two people who shake hands with the same number of people.

Solution. Each person's "handshake number" – the number of other persons he or she shook hands with – must be in the range $0, 1, \ldots, n-1$. The people are the pigeons, and the handshake numbers are the holes.

Is it possible that every hole is occupied? That would mean, in particular, that at least one person (say, Alice) shook hands 0 times, but at least one person (say, Bob) shook hands with everyone else at the conference. But Bob didn't shake hands with Alice! This is a contradiction, so at least one hole must be unoccupied.

Now there are n pigeons and $n-1$ remaining holes, so there must be two pigeons in one hole. That means that at least two people have the same handshake numbers, and we are done. □

Example 54. Given any 5 distinct points on the surface of a sphere show there exists a closed hemisphere that contains at least 4 of them.

(2002 Putnam)

Solution. Pick any two of the five points; say P and Q. There exists a great circle $\mathcal{C}$ through these two points, and the circle $\mathcal{C}$ is the boundary of two closed hemispheres. By the Pigeonhole Principle, one of these hemispheres must contain at least two of the remaining three points. Since it is closed, this hemisphere also contains P and Q, and thus contains at least four of the five given points. □

The following two examples seem very similar in their problem statements, but the sets we end up using as our "holes" are very different.

Example 55. Show that in any subset of 51 numbers taken from $\{1, 2, \ldots, 100\}$, there exists a pair of elements which do not have a common prime divisor.

Solution. Two consecutive integers cannot share a common prime divisor, since that prime would also have to divide the difference of the two numbers, namely 1. Consider the pairs of numbers $\{1,2\}, \{3,4\}, \{5,6\}, \ldots, \{99,100\}$. Each pair consists of two consecutive integers, and there are 50 pairs. By the Pigeonhole Principle, if we choose 51 numbers from $\{1, 2, \ldots, 100\}$, we must choose both numbers from at least one of the pairs. Such a pair gives us our two elements with no common prime divisor. □

Example 56. Show that in any subset of 51 numbers taken from $\{1, 2, \ldots, 100\}$, there exists a pair of elements such that one divides the other.

Solution. We define 50 subsets $S_1, S_2, S_3, \ldots, S_{50}$ of $\{1, 2, \ldots, 100\}$ in the following way. The smallest element of S_k is the odd number $2k-1$. In addition, S_k contains every number of the form $(2k-1)2^j$ that does not exceed 100. For example, $S_1 = \{1, 2, 4, 8, 16, 32, 64\}$, $S_{11} = \{21, 42, 84\}$, $S_{32} = \{63\}$. Now every number from 1 to 100 is included in some S_k. Also note that for a pair of numbers a, b both in a particular S_k with $a < b$, we have $a|b$.

Consider a subset of 51 numbers from $\{1, 2, \ldots, 100\}$. Since we have 50 subsets $S_1, \ldots, S_{50}$, by the Pigeonhole Principle there must exist some k ($1 \leq k \leq 50$) such that at least two elements of S_k are contained in our subset. In this pair of elements, one divides the other. □

The Pigeonhole Principle can often be useful in problems involving divisibility by a particular number n. We can take advantage of the facts that there are only n possible remainders upon division by n, and only $n-1$ of these are nonzero.

Example 57. Let $n \geq 1$ be a positive integer. If $a_1, \ldots, a_n$ are positive integers, prove that it is possible to paint some of these numbers green in such a way that the sum of the green numbers is divisible by n.

Solution. Define a new set of numbers $A_0, A_1, A_2, \ldots, A_n$ where $A_k = \sum_{i=1}^{k} a_i$ (so $A_0 = 0$, $A_1 = a_1$, $A_2 = a_1 + a_2$, $A_3 = a_1 + a_2 + a_3$, and so on).

Consider the remainders of the A_k upon division by n. We have n possible remainders for these $n+1$ divisions, so by the Pigeonhole Principle at least two must have the same remainder. We can take the difference of these numbers (say A_j and A_k with $j < k$) to get $A_k - A_j = a_{j+1} + a_{j+2} + \cdots + a_{k-1} + a_k$ which is a sum of a subset of the original numbers and is divisible by n. We paint $a_{j+1}, a_{j+2}, \ldots, a_k$ green, and we are done. □

Another interesting case occurs when we have infinitely many "pigeons" but only finitely many "holes." In this case, some hole must contain an infinite number of pigeons (otherwise, every hole would contain finitely many pigeons and the sum of finitely many finite numbers is finite, a contradiction). We apply this in the following example.

Example 58. Show that there is some four-digit sequence that occurs infinitely often as the first four digits of the powers of 2.

Solution. We know that there are $9 \cdot 10 \cdot 10 \cdot 10 = 9000$ possible four-digit sequences (excluding those that start with a zero). Imagine creating a bucket for each four-digit sequence, and sorting powers of 2 into the buckets according to their first four digits. (The first nine powers don't go in any bucket, but we have plenty more!) Since there are an infinite number of powers of 2, some bucket must contain infinitely many of them. The four-digit sequence corresponding to this bucket occurs infinitely often as the first four digits of powers of 2. □

Example 59. Two permutations $a_1, a_2, \ldots, a_{2010}$ and $b_1, b_2, \ldots, b_{2010}$ of the numbers $1, 2, \ldots, 2010$ are said to *intersect* if $a_k = b_k$ for some value of k in the range $1 \leq k \leq 2010$. Show that there exist 1006 permutations of the numbers $1, 2, \ldots, 2010$ such that any other permutation is guaranteed to intersect at least one of these 1006 permutations.

(2010 USAJMO)

Solution. Consider the set of permutations defined as follows: For each permutation $a_1, a_2, \ldots, a_{2010}$, we set $a_k = k$ for $1007 \leq k \leq 2010$ (i.e., all numbers greater than or equal to 1007 are fixed). For the ith permutation in our set, a_1 is set to be i. Each following element in the permutation will be the previous element plus one, with the exception that once we reach 1006 we loop back around to 1. This gives us the set of permutations

$$1, 2, 3, \ldots, 1006, 1007, 1008, \ldots, 2010$$
$$2, 3, 4, \ldots, 1006, 1, 1007, 1008, \ldots, 2010$$
$$3, 4, 5, \ldots, 1006, 1, 2, 1007, 1008, \ldots, 2010$$
$$\vdots$$
$$1005, 1006, 1, 2, \ldots, 1004, 1007, 1008 \ldots, 2010$$
$$1006, 1, 2, \ldots, 1004, 1005, 1007, 1008 \ldots, 2010$$

Note that our set of permutations has been structured specifically so that each number $1, 2, \ldots, 1006$ appears as the kth entry in one of our permutations for $1 \leq k \leq 1006$.

Consider any permutation $a_1, a_2, \ldots, a_{2010}$. There are 1006 entries in the range $a_1, a_2, \ldots, a_{1006}$, but only 1004 values above 1006. Since no value can occur twice in a permutation, one of these entries must take a value not exceeding 1006—say $a_k = j$, with $1 \leq k \leq 1006$ and $0 \leq j \leq 1006$. But then, as noted above, there is some permutation $b_1, \ldots, b_{2010}$ in our list such that $a_k = b_k = j$. Thus any permutation will intersect some permutation in our list, and we have shown the existence of 1006 permutations such that any other permutation is guaranteed to intersect at least one of these 1006 permutations. $\square$

One final technique to keep in mind is that in some cases, we can repeatedly apply the Pigeonhole Principle to obtain the desired result.

Example 60. Given a set M of 1985 distinct positive integers, none of which has a prime divisor greater than 23, prove that M contains a subset of 4 elements whose product is the 4th power of an integer.

(1985 IMO)

Solution. If no element in M has a prime divisor greater than 23, every element can be written in the form

$$2^{\alpha_1} 3^{\alpha_2} 5^{\alpha_3} 7^{\alpha_4} 11^{\alpha_5} 13^{\alpha_6} 17^{\alpha_7} 19^{\alpha_8} 23^{\alpha_9}$$

for some integers $\alpha_i \geq 0$. Notice that if two numbers have the same parities for the exponent of each prime, their product will be a perfect square. Since there are 2 possible parities (even or odd) for each of the 9 exponents, there are $2^9 = 512$ different parity distributions.

By the Pigeonhole Principle, since we have 1985 distinct integers and 512 parity distributions, there must be at least two numbers with the same parity distribution. We will set this pair aside, leaving us with 1983 distinct integers. We can apply the Pigeonhole Principle again to find another pair of numbers with the same parity distribution (not necessarily the same as the first set of numbers' distribution). Note that we can repeat this process 513 times (at which point we will have removed 1026 of the 1985 numbers, leaving us with 959 additional numbers; since this is still more than 512, we are fine applying the Pigeonhole Principle in this case).

Now we have 513 pairs of numbers such that each pair of numbers multiplies to a perfect square. Consider these 513 perfect squares. We will examine the remainders of the exponents in the prime factorization of these numbers upon division by 4. Since these are perfect squares, the only possible remainders are 0 and 2. Note that, similar to the parity case before, if two perfect squares have the same remainder distribution, their product will be a 4th power of an integer. Since we have 513 perfect squares and $2^9 = 512$ possible remainder distributions, by the Pigeonhole Principle there must be two perfect squares with the same remainder distribution.

Taking the numbers from our original set M that multiplied to the perfect squares which we multiplied to obtain our perfect 4th power, we have a subset of 4 elements whose product is the 4th power of an integer. $\square$

Chapter 8

Induction

Induction is a mathematical technique used to prove a claim is true for all integers greater than or equal to some fixed value (often for all nonnegative or all positive integers). When thinking about how a proof by induction works, it is often useful to imagine knocking over a line of dominoes; the basis for this metaphor will become clear as we investigate the proof procedure in more detail.

There are three key parts to a proof by induction: the base case, the induction hypothesis, and the inductive step. Suppose we have some statement we wish to prove for all n greater than or equal to some value.

- Base Case: For our base case, we show our statement is true for the smallest n value we are interested in (i.e., show we can knock over the first domino).
  - The value of n we use for our base case is a specific number.
  - For most problems, your base case will be for $n = 0$ or $n = 1$. However, if you are told to prove something for $n \geq 6$, then your base case could be $n = 6$.
  - Sometimes, it may be useful or even necessary to have more than one base case. Make sure you have enough in your base case to build on, but try to avoid creating more base cases than you need. Doing extra casework defeats the point of induction!

- Induction Hypothesis: The induction hypothesis is the assumption that our claim is true for some positive integer $n = k$, where k is some integer greater than or equal to our base case value (i.e., assume we know we can knock over the kth domino).
  - Why is this not assuming what we want to prove? There is a subtle but crucial distinction here between our statement we are trying

to prove and our induction hypothesis, namely for which values of n we assert our claim is true. The induction hypothesis makes a claim about a single value of n; we do not want to specify the exact numerical value here, so we use k as a placeholder. Our statement, on the other hand, is the assertion that our claim is true for *every* integer value of n greater than or equal to our base case value.

This distinction is why the inductive step is important. If we stopped with our induction hypothesis, we wouldn't actually have proven anything.

– Be careful not to assume k has any special properties either here or in the inductive step. Assumptions like this result in you only proving the statement for values of n that have those special properties, not all possible n values.

- Inductive Step: We now show that if the induction hypothesis is true (i.e., the claim holds for k), then the statement must be true for $n = k + 1$ as well (i.e., the dominos are arranged such that if the kth domino is knocked over, it knocks over the $(k + 1)$st domino).

– Start with the $n = k + 1$ case and break it down into smaller parts so you can use your induction hypothesis to prove it correct. Though it may be tempting, avoid starting with the case for $n = k$ and trying to build up to the $n = k + 1$ case. For some problems building up from the smaller case will work, but in other scenarios it can cause you to miss subcases. Starting with a general instance of the $n = k+1$ case and then reducing it to a smaller size problem will ensure you have covered everything.

– Make sure you do not assume the $n = k + 1$ case is true. If you do, you are assuming what you are trying to prove instead of actually proving it! How do we avoid this? As an example, if you are working with an equation, start with one side of the equation and manipulate the expression through a series of equalities to obtain the expression on the other side of your equation. Somewhere in you manipulations you will make use of your induction hypothesis.

Once we have completed these three parts, we're done! We have shown that our claim holds for some base value (for sake of example, let's say we have shown it is true for $n = 1$). We also have shown that if our claim holds for some generic value $n = k$, it must hold for $n = k + 1$ as well. This means since our claim holds for $n = 1$, it holds for $n = 1 + 1 = 2$. But then since it holds for $n = 2$, it must also hold for $n = 2 + 1 = 3$. And since it holds for $n = 3$, it also holds for $n = 3 + 1 = 4$ etc. This is why it is important we do not assume anything about the specific value of k in our induction hypothesis or inductive step; leaving k as a generic placeholder variable allows us to set up this domino chain effect.

If you get stuck when approaching an induction problem, try plugging in small values in place of n and look for a pattern to see what's going on. This is always a good starting point for problem solving, but it applies especially well to induction proofs!

To illustrate how each part of the proof works in practice, we examine the following well known identity.

Example 61. Prove that $\sum_{i=1}^{n} i = 1 + 2 + 3 + \cdots + n = \dfrac{n(n+1)}{2}$.

Solution. There are multiple ways to prove this result; here we proceed by induction on n.

Base Case: ($n = 1$). Plugging in, we see

$$\sum_{i=1}^{1} i = 1 = \frac{1 \cdot 2}{2}$$

so our base case holds.

Induction Hypothesis: Suppose the claim holds when $n = k$; that is,

$$\sum_{i=1}^{k} i = 1 + 2 + 3 + \cdots + k = \frac{k(k+1)}{2} \text{ for some } k \geq 1.$$

Inductive Step: Consider the left hand side of the equation when $n = k+1$. Our expression becomes $\sum_{i=1}^{k+1} i$. Note that

$$\sum_{i=1}^{k+1} i = (k+1) + \sum_{i=1}^{k} i.$$

By our induction hypothesis, we know that

$$\sum_{i=1}^{k} i = \frac{k(k+1)}{2}.$$

Substituting this in, we have

$$\sum_{i=1}^{k+1} i = k+1 + \frac{k(k+1)}{2} = \frac{2k+2}{2} + \frac{k^2+k}{2}$$

$$= \frac{k^2+3k+2}{2} = \frac{(k+1)(k+2)}{2}$$

$$= \frac{(k+1)((k+1)+1)}{2}$$

Since we have shown $\sum_{i=1}^{k+1} i = \frac{(k+1)(k+2)}{2}$, we know that if the identity holds for $n = k$, it holds for $n = k+1$ as well. By the mathematical principle of induction, this concludes our proof. □

We can also apply induction to prove a result we've seen earlier:

Example 62. Prove the Binomial Theorem for $n \geq 1$ using induction

$$(x+y)^n = \sum_{j=0}^{n} \binom{n}{j} x^j y^{n-j}.$$

Solution. We proceed by induction on n.
Base Case: ($n = 1$) We have

$$(x+y)^1 = x+y = \binom{1}{0} x^0 y^1 + \binom{1}{1} x^1 y^0,$$

so our base case holds.
Induction Hypothesis: Suppose for some $k \geq 1$, our claim holds when $n = k$; that is, we have

$$(x+y)^k = \sum_{j=0}^{k} \binom{k}{j} x^j y^{k-j}.$$

Inductive Step: Consider the left hand side of our equation when $n = k+1$. We have

$$(x+y)^{k+1} = (x+y)(x+y)^k = (x+y) \sum_{j=0}^{k} \binom{k}{j} x^j y^{k-j}$$

by our induction hypothesis. Expanding and rearranging, we have

$$(x+y)^{k+1} = \sum_{j=0}^{k} \binom{k}{j} x^{j+1} y^{k-j} + \sum_{j=0}^{k} \binom{k}{j} x^{j} y^{k-j+1}$$

$$= \sum_{i=1}^{k+1} \binom{k}{i-1} x^{i} y^{n-i+1} + \sum_{j=0}^{k} \binom{k}{j} x^{j} y^{k-j+1}$$

$$= \binom{k}{0} x^{0} y^{k+1} + \binom{k}{k} x^{k+1} y^{0} + \sum_{j=1}^{k} \left[\binom{k}{j} + \binom{k}{j-1} \right] x^{j} y^{k-j+1}$$

Using Pascal's Identity and the fact that

$$\binom{k}{0} = \binom{k+1}{0} = \binom{k}{k} = \binom{k+1}{k+1} = 1,$$

we simplify to yield

$$(x+y)^{k+1} = \binom{k+1}{0} x^{0} y^{k+1} + \binom{k+1}{k+1} x^{k+1} y^{0} + \sum_{j=1}^{k} \binom{k+1}{j} x^{j} y^{k-j+1}$$

$$= \sum_{j=0}^{k+1} \binom{k+1}{j} x^{j} y^{k-j+1}$$

as desired. By the principle of mathematical induction, this concludes our proof. □

Example 63. For all $n > 1$ and all sets $B, A_1, A_2, A_3, \ldots, A_n$, show

$$(A_1 \cap A_2 \cap \cdots \cap A_n) \cup B = (A_1 \cup B) \cap (A_2 \cup B) \cap (A_3 \cup B) \cap \cdots \cap (A_n \cup B).$$

Solution. <u>Base Case:</u> ($n = 2$). We wish to show that

$$(A_1 \cap A_2) \cup B = (A_1 \cup B) \cap (A_2 \cup B).$$

We can do this directly by shading the corresponding regions of a Venn Diagram and noting that the result is the same for both sides of the equation.

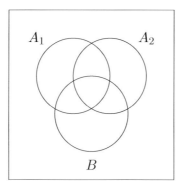

Induction Hypothesis: Suppose that when $n = k$ for some $k \geq 2$, our claim holds; that is, for any sets $B, A_1, A_2, \ldots, A_k$, we have

$$(A_1 \cap A_2 \cap \cdots \cap A_k) \cup B = (A_1 \cup B) \cap (A_2 \cup B) \cap (A_3 \cup B) \cap \cdots \cap (A_k \cup B).$$

Inductive Step: Consider the left hand side of our equation when $n = k + 1$.

$$(A_1 \cap A_2 \cap \cdots \cap A_k \cap A_{k+1}) \cup B.$$

Let $C = A_1 \cap A_2 \cap \cdots \cap A_k$. Then our expression becomes $(C \cap A_{k+1}) \cup B$. We know by our base case that this equals

$$(C \cup B) \cap (A_{k+1} \cup B) = [(A_1 \cap A_2 \cap \cdots \cap A_k) \cup B] \cap (A_{k+1} \cup B).$$

Now we can apply our Induction Hypothesis to our first term to yield

$$[(A_1 \cup B) \cap (A_2 \cup B) \cap (A_3 \cup B) \cap \cdots \cap (A_k \cup B)] \cap (A_{k+1} \cup B)$$

which is the same as

$$(A_1 \cup B) \cap (A_2 \cup B) \cap (A_3 \cup B) \cap \cdots \cap (A_k \cup B) \cap (A_{k+1} \cup B)$$

as desired. Since we have not made any assumptions about our sets, we know our claim holds for any sets $B, A_1, \ldots, A_k, A_{k+1}$. By the mathematical principle of induction, this concludes our proof. □

Induction can also be applied when the claim we want to prove is not an equation. Let's look at some examples.

Example 64. Prove that for any $n \geq 1$, a $2^n \times 2^n$ board with any one unit cell removed can be exactly covered with non-overlapping L-shaped trominoes.

Solution. We proceed by induction on n.
Base Case: ($n = 1$) We have a 2×2 grid with one square removed. Regardless of which cell is removed, we are left with exactly 3 squares in an L-shape, which can be covered by a single tromino.
Induction Hypothesis: Suppose that our claim holds when $n = k$ for some $k \geq 1$; that is, any $2^k \times 2^k$ grid with a single square removed can be tiled using L-shaped trominoes.
Inductive Step: Consider the board when $n = k + 1$: a $2^{k+1} \times 2^{k+1}$ grid with one square removed. We can divide this grid into four equally sized quadrants, each $2^k \times 2^k$. One of these quadrants contains the missing square; by our induction hypothesis, this quadrant can be tiled using L-shaped trominos. We place one tromino in the middle of our grid such that it covers a corner square of each of the remaining three quadrants. We can treat these covered cells

as being removed from those quadrants, leaving us with three $2^k \times 2^k$ grids with one square removed from each. By our induction hypothesis, we can tile each of these with L-shaped trominoes, so we have covered the entire grid with these trominoes. Thus, any $2^{k+1} \times 2^{k+1}$ grid with a single cell removed can be exactly covered using L-shaped trominoes. By the principle of mathematical induction, this concludes our proof. □

Sometimes we may want to prove a statement true only for a certain infinite subset of the integers (eg. positive odd integers, only integers with remainder 1 upon division by 4, etc.). In such a case, we write the desired value as a function of the variable we induct on and proceed from there. In the following example, we wish to prove a statement only for odd sized groups.

Example 65. Suppose that $n \geq 3$ children are playing on a playground, each holding a custard pie. Suddenly every child throws his or her pie at the face of the child that is closest to them. Assume that all the distances between the children are distinct so this rule is well defined. Prove, by induction, that if n is odd, then at least one child has no pie thrown at him or her.

Solution. We let $n = 2k - 1$ to ensure we only look at odd n. We induct on k.
Base Case: ($k = 1$). Note that the two children with the closest distance between them will throw their pies at each other. The last child throws her pie at one of those two, but is not pied. Thus our claim holds.
Induction Hypothesis: Suppose that for a group of $n = 2k - 1$ children for some $k \geq 1$, any arrangement results in at least one child not being pied.
Inductive Step: Consider the case of $n = 2(k+1) - 1 = 2k + 1$ children. Let Alice and Bob be the two children with the shortest distance between them. Then Alice throws her pie at Bob and Bob throws his pie at Alice. Ignoring Alice and Bob, we know by our induction hypothesis that there must be some child Joe amongst the other $2k - 1$ children who is not pied. This child also is not pied by Alice or Bob, and adding Alice and Bob back in will not cause another child to switch targets to Joe (a child could only switch his or her target to Alice or Bob). Thus Joe has no pie thrown at him. By the mathematical principle of induction, this concludes our proof. □

Example 66. A group of n people play a round-robin tournament. Each game ends in either a win or loss (there are no ties). Show that it is possible to label the players $P_1, P_2, P_3, \ldots, P_n$ in such a way that P_1 defeated P_2, P_2 defeated P_3, ..., and P_{n-1} defeated P_n.

Solution. We proceed by induction on n.
Base Case: ($n = 2$) In a two person tournament, one game is played with one player winning and the other losing. The winning player is labelled P_1 and

the other player is labelled P_2. This yields an ordering matching our criteria, so our base case holds.

Induction Hypothesis: Suppose that when $n = k$ for some $k \geq 2$, in any round-robin tournament with k players we can label the players $P_1, P_2, P_3, \ldots, P_k$ in such a way that P_1 defeated P_2, P_2 defeated P_3, ..., and P_{k-1} defeated P_k.

Inductive Step: Consider a round-robin tournament with $k+1$ players. We will pick one player (Player X), and temporarily ignore him. Doing so, we are left with k players to consider, and by our induction hypothesis, there exists some ordering $P'_1, P'_2, \ldots, P'_k$ such that P'_1 defeated $P'_2, \ldots, P'_{k-1}$ defeated P'_k. We need to find somewhere in this ordering to insert Player X; if we can find such a placement, we will have an ordering of all $k+1$ players meeting our criteria.

If Player X was undefeated, we can take the ordering $X, P'_1, P'_2 \ldots, P'_n$ and label the players accordingly. Otherwise, Player X lost to at least one other player. Let j be the largest index $1 \leq j \leq n$ such that P'_j defeated Player X. Since j is the largest such index, Player X must have won against P'_{j+1} (or $j = k$), and thus $P'_1, \ldots, P'_j, X, P'_{j+1}, \ldots, P'_k$ is a valid ordering for our criteria. We can label the players in this order accordingly.

By the principle of mathematical induction, this concludes our proof. $\square$

Example 67. A group of n chickens are having a tournament where every chicken plays one game against every other competing chicken (life as a chicken gets boring). Every match ends in a win or loss for the players (i.e. there are no ties), the winner being the chicken who pecks the other first. We say a chicken u virtually pecks another chicken v if u pecks some other chicken w who pecked v. A chicken is called a King Chicken if, when the tournament ends, that chicken either pecked or virtually pecked every other chicken in the tournament. Prove that every tournament must have at least one King Chicken.

Solution. We prove this by induction on n.

Base Case: ($n = 2$) The smallest reasonable tournament is one with two chickens. One chicken must peck the other; this is the King Chicken. Thus any tournament with two chickens has a King.

Induction Hypothesis: Suppose that for $n = k$, our claim holds; that is, any tournament of k chickens for some $k \geq 2$ has a King Chicken.

Inductive Step: Consider a tournament with $k+1$ chickens. We will temporarily set aside one of them (say Chicken Little) and focus on the remaining k chickens. By our induction hypothesis, there must be some chicken among those k chickens that is a King for the sub-tournament of just those k chickens. Now we consider Chicken Little.

If Chicken Little is pecked by the King of the sub-tournament, that King is a King Chicken for the entire tournament.

If Chicken Little is pecked by some chicken that the King of the sub-tournament pecked, the King Chicken virtually pecks Chicken Little and thus is a King Chicken for the entire tournament.

The remaining case is where Chicken Little pecks the King of the sub-tournament but is not pecked by any chicken pecked by that King (i.e. Chicken Little pecks all the chickens pecked by the King of the sub-tournament). We claim that Chicken Little is a King Chicken in this case. Note that any chicken not pecked by Chicken Little was not pecked by the King of the sub-tournament either. Thus this chicken must have been pecked by some chicken the King of the sub-tournament pecked. Since Chicken Little pecks all chickens the King of the sub-tournament pecked, this implies Chicken Little virtually pecks this chicken, and thus Chicken Little pecks or virtually pecks all other chickens.

By the principle of mathematical induction, this concludes our proof. □

Strong Induction

There are many variations on induction; we will discuss one common one ("strong induction") here.

- How is strong induction different from "normal" induction?

    - The only change is in our induction hypothesis. Instead of assuming that our claim is true for $n = k$ with k being some integer greater than or equal to our base case value, we suppose that our claim is true for every integer greater than or equal to our base case value, but less than or equal to k.

      For example, if our base case is $n = 1$, our induction hypothesis would suppose the claim is true for $n = 1$, $n = 2$, $n = 3$, ..., $n = k$.

- Why is this proof method still valid?

    - At first glance, it seems like we're assuming *a lot* here. However, when we break down our induction hypothesis, we see that there is the exact same domino effect as with normal induction. We still start with our base case (say $n = 1$). Next, we prove that if our claim holds true for every integer n with $1 \leq n \leq k$, it holds true for $n = k + 1$.

      Now we build up from our base case. Since our claim holds for every integer n with $1 \leq n \leq 1$, it must hold for $n = 1 + 1 = 2$. Since it holds for $n = 1$ and $n = 2$, it holds for every integer $1 \leq n \leq 2$ and thus must also hold for 3. This pattern continues on forever.

- How do we know when to use strong induction versus normal induction?

– There's no definitive rule that governs this, but you should generally default to normal induction. If you get to your inductive step (the $n = k+1$ case) and find that you can't figure out a way to reduce to the $n = k$ case, but you *can* figure out a way to reduce to a case (or multiple cases) for smaller values of n, it's likely that strengthening your assumption and taking advantage of strong induction will help.

One typical application of strong induction is proving closed forms for sequences based on their recurrence relations. If you are not familiar with recurrence relations, the next section explores them in detail.

Example 68. Prove Binet's formula gives a closed form for the Fibonacci numbers:

$$F_n = \frac{1}{\sqrt{5}} \left[\left(\frac{1+\sqrt{5}}{2}\right)^n - \left(\frac{1-\sqrt{5}}{2}\right)^n \right]$$

Recall that the Fibonacci numbers are defined by $F_0 = 0$, $F_1 = 1$ and for $n \geq 2$, $F_n = F_{n-1} + F_{n-2}$.

Solution. <u>Base Case</u>: In order to use our recurrence, we will need two base cases: $n = 0$ and $n = 1$. We have

$$\frac{1}{\sqrt{5}} \left[\left(\frac{1+\sqrt{5}}{2}\right)^0 - \left(\frac{1-\sqrt{5}}{2}\right)^0 \right] = \frac{1}{\sqrt{5}}(1-1) = 0 = F_0$$

and

$$\frac{1}{\sqrt{5}} \left[\left(\frac{1+\sqrt{5}}{2}\right)^1 - \left(\frac{1-\sqrt{5}}{2}\right)^1 \right] = \frac{1}{\sqrt{5}}(\sqrt{5}) = 1 = F_1$$

so our base cases hold.

<u>Induction Hypothesis</u>: Suppose that for all n such that $0 \leq n \leq k-1$, we know that Binet's formula gives us the nth Fibonacci number.

<u>Inductive Step</u>: Consider F_k. We know by our recurrence that

$$F_k = F_{k-1} + F_{k-2},$$

Induction

and by our strong induction hypothesis we can apply Binet's formula to get

$$F_k = F_{k-1} + F_{k-2}$$
$$= \frac{1}{\sqrt{5}}\left[\left(\frac{1+\sqrt{5}}{2}\right)^{k-1} - \left(\frac{1-\sqrt{5}}{2}\right)^{k-1}\right]$$
$$+ \frac{1}{\sqrt{5}}\left[\left(\frac{1+\sqrt{5}}{2}\right)^{k-2} - \left(\frac{1-\sqrt{5}}{2}\right)^{k-2}\right]$$
$$= \frac{1}{\sqrt{5}}\left[\left(\frac{1+\sqrt{5}}{2}\right)^{k-2}\left(\frac{3+\sqrt{5}}{2}\right) - \left(\frac{1-\sqrt{5}}{2}\right)^{k-2}\left(\frac{3-\sqrt{5}}{2}\right)\right]$$

Note that

$$\left(\frac{1+\sqrt{5}}{2}\right)^2 = \frac{3+\sqrt{5}}{2} \quad \text{and} \quad \left(\frac{1-\sqrt{5}}{2}\right)^2 = \frac{3-\sqrt{5}}{2}$$

so this simplifies to

$$F_k = \frac{1}{\sqrt{5}}\left[\left(\frac{1+\sqrt{5}}{2}\right)^k - \left(\frac{1-\sqrt{5}}{2}\right)^k\right]$$

as desired. By the principle of mathematical induction, this concludes our proof. $\square$

Example 69. Prove that every positive integer n has a binary representation. That is, show that every positive integer n can be expressed as

$$n = c_j 2^j + c_{j-1} 2^{j-1} + \cdots + c_2 2^2 + c_1 2^1 + c_0 2^0$$

where each of the c_i is either 0 or 1.

Solution. We proceed by strong strong induction on n.
Base Case: ($n = 1$) The number 1 is simply $1 \cdot 2^0$, so it has a binary representation.
Induction Hypothesis: Suppose that for all n such that $1 \leq n \leq k-1$, n has a binary representation.
Inductive Step: We consider two cases for $n = k$ based on its parity. If k is odd, then $k-1$ is even, so $\frac{k-1}{2}$ is a positive integer less than k and by our induction hypothesis it must have some binary representation

$$\frac{k-1}{2} = c_j 2^j + c_{j-1} 2^{j-1} + \cdots + c_2 2^2 + c_1 2^1 + c_0 \cdot 2^0.$$

Multiplying by 2 gives us a binary representation for $k-1$:
$$k-1 = c_j 2^{j+1} + c_{j-1} 2^j + \cdots + c_2 2^3 + c_1 2^2 + c_0 \cdot 2^1.$$

Adding 1 yields our binary representation for k:
$$k = c_j 2^{j+1} + c_{j-1} 2^j + \cdots + c_2 2^3 + c_1 2^2 + c_0 \cdot 2^1 + 1 \cdot 2^0.$$

Suppose on the other hand that k is even. Note that $\frac{k}{2}$ is a positive integer less than k, so we know by our induction hypothesis that it has some binary representation
$$\frac{k}{2} = c_j 2^j + c_{j-1} 2^{j-1} + \cdots + c_2 2^2 + c_1 2^1 + c_0 2^0.$$

Multiplying through by 2, we get
$$k = c_j 2^{j+1} + c_{j-1} 2^j + \cdots + c_2 2^3 + c_1 2^2 + c_0 2^1,$$

which is a binary representation for k. Since every positive integer is either odd or even, by the principle of mathematical induction this concludes our proof. □

Chapter 9

Recurrence Relations

Although the Rules of Sum and Product provide powerful tools for counting, there are problems which cannot be solved easily with just these. Some problems resist solutions using more direct methods and instead they can be solved easily using recurrence relations. In this sort of problem, we have a general count that we want to do which depends on a free parameter, say n. For instance, returning to Example 5, we could want to count the number of subsets of $\{1, 2, \ldots, n\}$. One approach to solving such a problem is to look at the sequence of all such counts as n varies. In the case of Example 5, we would let $(a_n)_{n \geq 0}$ be the sequence with a_n equal to the number of subsets of $\{1, 2, \ldots, n\}$. We know from Example 5 that $a_n = 2^n$, but pretend that you do not know this for the moment. Instead of looking for a formula for a_n, we look for a formula for a_n in terms of all the previous terms in the sequence, $a_1, \ldots, a_{n-1}$. In the case of Example 5, this is very easy. A subset of $\{1, 2, \ldots, n\}$ either contains n or it doesn't. If it contains n, then it is built from a subset of $\{1, 2, \ldots, n-1\}$ by adding n. If it doesn't, then it is already a subset of $\{1, 2, \ldots, n-1\}$. Hence, we see that there are a_{n-1} subsets of $\{1, 2, \ldots, n\}$ that contain n and another a_{n-1} that do not contain n. Thus $a_n = 2a_{n-1}$.

There is an analogy between recurrence relations and induction. In the inductive step of a (strong) induction proof, we show that if we can prove all previous cases of the induction hypothesis, then we can prove the next one. A recurrence relation tells us that if we know all previous terms in the sequence (a_n), then we could compute the next one.

If we can find a recurrence relation for the sequence (a_n), then we can compute a_n reasonably efficiently (in roughly n steps). For some problems this is the best one can do. However, for many recurrence relations, we can use the recurrence to obtain an explicit formula for the general term a_n. For example, for the recursion $a_n = 2a_{n-1}$, we easily see that the general solution to this recurrence is $a_n = C2^n$ for some constant C. Using the value of any one term in the sequence (say $a_0 = 1$, since the only subset of the empty set is the empty set itself), we can determine that $C = 1$ and we again have $a_n = 2^n$. Example 68 in the previous section provides another example where we turned a recurrence relation for the Fibonacci numbers into a concrete formula. We also refer the reader to the appendix for some theoretical results on solving recurrence relations

Example 70. We are given sufficiently many tiles of the forms of a 2×1 rectangle and 1×1 square. Let $n > 3$ be a natural number. In how many ways can one tile a rectangle $3 \times n$ using these tiles, so that no two 2×1 rectangles have overlaps(their borders are adjacent), and each of them has the longer side parallel to the shorter side of the big rectangle?

(Austria 2003)

Solution. Call such a tiling with no two 2×1 rectangles having a common point a *good tiling*. Let us denote by a_n the number of good tillings of a $3 \times n$ rectangle. We will start from $n = 1$, insisting that all the 2×1 rectangles are oriented with the long side vertical. For $n > 3$ this agrees with the phrasing of the problem. Drawing the possible pictures, we see that there are 3 ways to tile a 3×1 rectangle, so $a_1 = 3$.

In the case $n = 2$ split the 3×2 rectangle into two 3×1 vertical bands. If in one of the bands we use a 2×1 rectangle, then we can use no such rectangle in the other band, so we must fill in the rest with squares. There are 4 possible placements for the rectangle in the two bands, hence four tilings that use a rectangle. Adding in the tiling using only 1×1 squares, we get $a_2 = 5$.

In the general situation, we extend this same approach. For a $3 \times (n+1)$ rectangle, we have $n+1$ vertical bands of length 3. If we place a 2×1 rectangle in the $(n+1)$-st band, then we are forced to not have a rectangle in the n-th band. Thus the n-th band is tiled using only 1×1 squares. Since there are two ways to place the 2×1 rectangle in the last band, and we can place any good tiling for a $3 \times (n-1)$ rectangle in the first $n-1$ bands, we see that there are $2a_{n-1}$ tillings of this form. On the other hand, if we do not place a 2×1 rectangle in the last band, then we must tile it in the only way possible with 1×1 squares. In the remaining n bands, we place an arbitrary good tiling for the $3 \times n$ rectangle. This gives us a_n possible tiling. Combining these cases, we obtain the recurrence $a_{n+1} = a_n + 2a_{n-1}$, for $n \geq 2$.

The theory of second-order constant coefficient linear recurrences lets us use this recurrence to write down a general formula for a_n. The characteristic polynomial for this linear recurrence is $r^2 - r - 2 = 0$ which has roots $r_1 = 2$ and $r_2 = -1$. Thus we see that the sequences 2^n and $(-1)^n$ satisfy the recurrence. Hence the general form for such a sequence is $a_n = C_1 2^n + C_2 (-1)^n$ for some constants C_1 and C_2. Plugging in the initial values $a_1 = 3$ and $a_2 = 5$ gives us two linear equations $2C_1 - C_2 = 3$ and $4C_1 + C_2 = 5$. Solving these gives $C_1 = 4/3$ and $C_2 = -1/3$, hence we obtain that

$$a_n = \frac{2^{n+2} + (-1)^{n+1}}{3}.$$

(We could have simplified the arithmetic slightly if we had begun the recursion from $a_0 = 1$.) □

Example 71. Find the number of subsets of $\{1, 2, \ldots, n\}$ that contain no two consecutive elements.

Solution. Let us denote the number of such sets by a_n. To obtain a recursion, we look at the last element n. If n is in our set, then $n - 1$ cannot be, so we are looking at similar subsets of $\{1, 2, \ldots, n - 2\}$. Note that to each such subset, we can adjoin n and get a set with no consecutive numbers. Thus the subsets with no consecutive elements containing n are in bijection with the subsets with no consecutive elements of $\{1, 2, \ldots, n - 2\}$, so there are a_{n-2} of these.

The other case is when n is not in our subset and then we have a subset of $\{1, 2, \ldots, n - 1\}$ with no consecutive elements. There are a_{n-1} of these.

Putting it all together, we get the recurrence relation $a_n = a_{n-1} + a_{n-2}$. Note that $a_1 = 2$, $a_2 = 3$, so we see that $a_n = F_{n+2}$ where F_n are the Fibonacci numbers. □

Example 72. Find the number of strings of n letters, each equal to A, B, C, D such that the same letter never occurs three times in succession.

Solution. Let us call us such strings *good*. We look for a recursion again. Look at what happens when we try to add a new letter to a string of length $n - 1$. There are usually 4 choices, but if the string ends in a double letter, then we cannot adjoin this letter again and there are only 3 choices.

The idea is to define two sequences. Let a_n be the number of *good* strings of length n and let b_n be the number of *good* strings of length n ending in a double letter. We will find two recursions relating a_n and b_n, respectively, to the earlier terms in both sequences. This is more work that finding a recursion for a single sequence, but it makes the individual recursions simpler. This in turn enables us to find more complicated recursions.

We almost found the recursion for a_n above. If a good string of length $n - 1$ does not end in a double letter (and there are $a_{n-1} - b_{n-1}$ such strings), then there are 4 ways we can add a letter to get a good string of length n. If it does end in a double letter, then there are only 3 ways to add a letter. Hence we get $a_n = 4(a_{n-1} - b_{n-1}) + 3b_{n-1} = 4a_{n-1} - b_{n-1}$.

The recursion for b_n is similar. If a good string of length $n-1$ does not end in a double letter, then by repeating its last letter we get a good string of length n which ends in a double letter. If a good string of length $n-1$ ends in a double letter, then there is no way to add a single letter to get a good string of length n ending in a double letter. Hence we get $b_n = a_{n-1} - b_{n-1}$.

Since we do not need to compute the sequence b_n, we can eliminate it from our equations to get a recursion just involving a_n. The first recurrence gives $b_{n-1} = 4a_{n-1} - a_n$ and hence $b_n = 4a_n - a_{n+1}$. Plugging these into the second recursion gives $4a_n - a_{n+1} = a_{n-1} - (4a_{n-1} - a_n)$ or $a_{n+1} = 3a_n + 3a_{n-1}$. Shifting the index we can write this as $a_n = 3a_{n-1} + 3a_{n-2}$ for $n \geq 3$. It is easy to compute the first two terms in the sequence ($a_1 = 4$ and $a_2 = 16$), so this would allow us to compute any a_n.

To get a closed formula for a_n, we look at the characteristic polynomial $r^2 - 3r - 3 = 0$ of the recursion. This polynomial has roots

$$r_1 = \frac{3 + \sqrt{21}}{2} \text{ and } r_2 = \frac{3 - \sqrt{21}}{2}.$$

Hence the general solution to this recursion is

$$a_n = C_1 \left(\frac{3 + \sqrt{21}}{2}\right)^n + C_2 \left(\frac{3 - \sqrt{21}}{2}\right)^n$$

for constants C_1 and C_2. Plugging in the first two terms in the sequence and solving the resulting equations for C_1 and C_2, we find after a little algebra that

$$a_n = \frac{4}{3\sqrt{21}} \left(\left(\frac{3 + \sqrt{21}}{2}\right)^{n+1} - \left(\frac{3 - \sqrt{21}}{2}\right)^{n+1} \right). \qquad \square$$

Example 73. We divide a circle into n circular sectors. Find the number of ways to color these sectors with m colors such that no two adjacent sectors have the same color.

Solution. Call such a coloring *good*. Let us denote by c_n the number of *good* colorings of the circle split up into n sectors.

Let's see what happens if carelessly we try to count colorings using only the Rule of Product. We can pick the color for the first sector in m ways, the color for the second one in $m-1$ ways, the next one again in $m-1$ ways, and continuing until the end we have $m-1$ colors for the last sector. This would give a total of $m \cdot (m-1)^{n-1}$ colorings.

The reader should notice immediately that this count is not correct. We have not ensured that the first sector and last sector have different colors.

Recurrence Relations

The remarkable thing is that if we think more carefully about what we have counted, then we will get a recursion for the sequence c_n.

If the first sector and the last sector have different colors, then we get a *good* coloring for n sectors. By definition, there are c_n such colorings. If we end up with the same color for the first and last sector, then we do not get a *good* coloring. However, in this case we note that we can fuse the first and last sector and obtain a *good* coloring for a circle divided into $n-1$ sectors. Thus there are c_{n-1} colorings of this type. Thus we have the recursion $c_n + c_{n-1} = m(m-1)^{n-1}$. The reader should be careful because this is only valid for $n \geq 3$. If we write down the recursion for $n=2$, we get $c_1 + c_2 = m(m-1)$. It is easy to count $c_2 = m(m-1)$ and $c_1 = m$ so this is an issue. We can introduce a false value for c_1, namely pretend it is 0 and the arithmetic will actually be easier to solve the given recurrence relation. To find a closed formula for c_n, let us look at the sum $\sum_{j=2}^{n}(-1)^{n-j}(c_j + c_{j-1})$. On one hand, if we unravel this sum by looking at cancellation between consecutive terms, it is equal to $c_n + (-1)^{n-2}c_1 = c_n$. On the other hand, by the above recurrence, it is equal to $m\sum_{j=2}^{n}(-1)^{n-j}(m-1)^{j-1}$. This sum is a geometric series, which we can sum to get

$$c_n = m \cdot \frac{(m-1)^n + (-1)^n(m-1)}{m-1+1} = (m-1)^n + (-1)^n(m-1). \quad \square$$

Example 74. Find the number of functions $f : \{1, 2, \ldots, n\} \to \{1, 2, 3, 4, 5\}$ such that $|f(k+1) - f(k)| \geq 3$ for all $k = 1, 2, \ldots, n-1$.

(Romania 2000)

Solution. The idea is pretty straightforward: we have to split up the problem according to the value of $f(n)$ and obtain recursions. There is one slight catch about which we must be careful. If $n = 1$, then it is easy to see that there are 5 such functions since $f(1)$ can have any value. However, if $n \geq 2$, then the function f can never take on the value 3, since it would contradict the inequality. This will make us start our recursion later than you might expect.

So, let a_n, b_n, c_n and d_n be the number of such functions with $f(n) = 1$, $f(n) = 2$, $f(n) = 4$, and $f(n) = 5$, respectively. We are interested in the total number of functions $S_n = a_n + b_n + c_n + d_n$, for $n \geq 2$. (Again, for $n = 1$, the total number of functions is $S_1 = 5$, but $a_1 + b_1 + c_1 + d_1 = 4$.)

For $f(n) = 1$, we have that the possibilities for $f(n-1)$ are 4 and 5, so $a_n = c_{n-1} + d_{n-1}$.

For $f(n) = 2$, we have that the only possibility is $f(n-1) = 5$, so $b_n = d_{n-1}$.

For $f(n) = 4$, again we are forced to have $f(n-1) = 1$, thus $c_n = a_{n-1}$.

Finally, if $f(n) = 5$, we have two valid values for $f(n-1)$, namely 1 and 2, and thus $d_n = a_{n-1} + b_{n-1}$.

Now, let us note that adding the first and the last, we have

$$a_n + d_n = a_{n-1} + b_{n-1} + c_{n-1} + d_{n-1} = S_{n-1}.$$

Adding the middle two, we have

$$b_n + c_n = a_{n-1} + d_{n-1} = S_{n-2}.$$

Finally, summing these up, we see that $S_n = S_{n-1} + S_{n-2}$ for $n \geq 4$ (to ensure that $n - 2 \geq 2$). Note that $S_2 = 6$ and $S_3 = 10$. Recognizing the Fibonacci recursion, we see that $S_n = 2F_{n+2}$ for $n \geq 2$. □

Example 75. A subset X of $\{1, 2, \ldots, n\}$ is *selfish* if $|X| \in X$ (for a set A, $|A|$ denotes the number of elements). Find the number of selfish sets with the property that they have no proper *selfish* subsets.

(Putnam 1996)

Solution. Again, it is a good idea, if we want to find a recurrence relation, to look at the last element n. It might be or not be in our minimal *selfish* set. Let a_n be the number of minimal selfish subsets containing n and b_n the number which do not contain n.

It is easy to see that those which do not contain n are minimal *selfish* subsets of $\{1, 2, \ldots, n-1\}$, so $b_n = a_{n-1} + b_{n-1}$.

Now consider a minimal *selfish* set that contains n. If $n > 1$, then such a set cannot contain 1 (since if it did $\{1\}$ would be a proper *selfish* subset). Thus the cardinality is not equal to n. Hence such a set must have at least two distinct elements, n and its cardinality.

Now we can do the following trick, remove n from the set and lower the rest of numbers in our *selfish* set by 1. Note that this cannot produce a 0, because 1 cannot be in our set. Also, this set has cardinality one less than the old cardinality. Hence, we obtain a *selfish* subset of $\{1, 2, \ldots, n-2\}$. We claim it is minimal.

This is true, since we can reverse the procedure. Namely, if Y is a proper *selfish* subset of the resulting set, then we add 1 to each element of Y and adjoin n. This would give rise to a proper *selfish* subset of our original set.

Thus we see that $a_n = a_{n-2} + b_{n-2}$. We are interested in the total number of minimal *selfish* sets, $c_n = a_n + b_n$. Putting together the relations we have, we get $c_n = c_{n-1} + c_{n-2}$.

We recognize this as the Fibonacci recursion. Since $c_1 = 1$ and $c_2 = 1$, we conclude that $c_n = F_n$, the n-th Fibonacci number. □

Recurrence Relations 85

Example 76. Find the number of strings of length n made up of the digits $\{0, 1, 2\}$ such that they contain none of the following strings 100, 101, 200 or 201.

Solution. Again, the idea is to look at what happens when when we try to extend a string of length $n - 1$. As usual, we call a string *good* if it does not contain any of the excluded substrings from the problem statement. Let s_n be the number of *good* strings of length n, let a_n be the number of *good* strings of length n that end in either a 1 or a 2, and let b_n the number of *good* strings that end in either 10 or 20. (Here by convention we have $a_0 = b_0 = b_1 = 0$.)

If a string of length at least 2 does not end in either a 1, 2, 10, or 20, then it must end in 00. Since the substrings 100 and 200 are excluded, this means it must end in 000 (or $n = 2$). Iterating this we see the string must be all zeroes. It is also easy to see that for $n = 0$ or 1 the only strings counted by s_n, but not by a_n or b_n, are the strings with only zeroes. Hence we see that $s_n = a_n + b_n + 1$ for all n.

We can always append a 2 to any *good* string of length $n - 1$ to obtain a *good* string of length n that ends in a 2. However, to get a *good* string of length n that ends in a 1, we can only append a 1 to a *good* string of length $n - 1$ that ends in either a 1 or a 2 or is the all zeroes string. Thus we see that $a_n = s_{n-1} + a_{n-1} + 1$.

Finally, a *good* string of length n that ends in 10 or 20 is clearly obtained from a *good* string of length $n - 1$ ending in 1 or 2 by appending a 0. Thus $b_n = a_{n-1}$.

It is easy to see that these three recursions (and the initial values above), would let us compute a_n, b_n, and s_n for all $n \geq 0$. To get a closed formula for s_n, we first eliminate a_n and b_n from the recursions. Eliminating b_n is easy. To eliminate a_n, add and subtract the first two recursions to get $s_n - s_{n-1} = 2a_{n-1} + 2$ and $s_n + s_{n-1} = 2a_n$. Shifting the index on the second recursion and using it to eliminate a_{n-1} gives $s_n - s_{n-1} = s_{n-1} + s_{n-2} + 2$ or $s_n = 2s_{n-1} + s_{n-2} + 2$.

This type of recursion is called an inhomogeneous because of the extra term not depending on the sequence s_n. To solve it, we note that $t_n = s_n + 1$ satisfies the homogeneous recursion $t_n = 2t_{n-1} + t_{n-2}$. (One way to find this is to guess that for some constant c, $t_n = s_n + c$ should satisfy a homogeneous recursion. Plugging in $s_n = t_n - c$ gives after some algebra $t_n = 2t_{n-1} + t_{n-2} + (2 - 2c)$, hence we conclude that this equation will be homogeneous if $c = 1$.)

To solve this recursion, we note that the characteristic polynomial

$$r^2 - 2r - 1 = 0$$

has roots $r_1 = 1 + \sqrt{2}$ and $r_2 = 1 - \sqrt{2}$. Hence

$$t_n = C_1(1 + \sqrt{2})^n + C_2(1 - \sqrt{2})^n.$$

From the initial conditions $t_0 = s_0 + 1 = 2$ and $t_1 = s_1 + 1 = 4$, we get $C_1 + C_2 = 2$ and $(1+\sqrt{2})C_1 + (1-\sqrt{2})C_2 = 4$. Solving these linear equations gives $C_1 = 1 + 1/\sqrt{2}$ and $C_2 = 1 - 1/\sqrt{2}$. So after a little rearranging we get

$$s_n = \frac{1}{\sqrt{2}}((1+\sqrt{2})^{n+1} - (1-\sqrt{2})^{n+1}) - 1. \qquad \square$$

Example 77. Let A and E be opposite vertices of a regular octagon. A frog starts at vertex A. From any vertex except E it jumps to one of the two adjacent vertices. When it reaches E it stops. Let a_n be the number of distinct paths of exactly n jumps ending at E. Prove that:

$$a_{2n-1} = 0, \quad a_{2n} = \frac{(2+\sqrt{2})^{n-1} - (2-\sqrt{2})^{n-1}}{\sqrt{2}}.$$

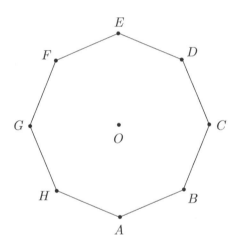

(IMO 1979)

Solution. The idea is again to obtain recurrence. Let a_n, b_n, c_n, d_n, e_n, f_n, g_n and h_n the number of distinct path of length n ending at A, B, C, D, E, F, G and H respectively. We need to obtain e_n.

We know by symmetry of the figure that $b_n = h_n$, $c_n = g_n$ and $d_n = f_n$; pairing vertices that have the same distance to A, the starting point. Now it is also easy to see that

$$e_n = d_{n-1} + f_{n-1} = 2d_{n-1}$$
$$d_n = c_{n-1}$$
$$a_n = 2b_{n-1}$$
$$c_n = b_{n-1} + d_{n-1}$$
$$b_n = a_{n-1} + c_{n-1}$$

From the first two we obtain $d_n = \dfrac{e_{n+1}}{2}$, $c_n = \dfrac{e_{n+2}}{2}$. Now substitute into the fourth to get $b_{n-1} = \dfrac{e_{n+2} - e_n}{2}$. From the third equation we have $a_n = e_{n+2} - e_n$.

Putting it together in the fifth we obtain the recurrence relation

$$e_{n+3} - 4e_{n+1} + 2e_{n-1} = 0$$

We thus obtain the characteristic equation $x^4 - 4x^2 + 2 = 0$ which has roots $\pm\sqrt{2 \pm \sqrt{2}}$. We thus have that

$$e_n = (A + (-1)^n B)\left(\sqrt{2+\sqrt{2}}\right)^n + (C + (-1)^n D)\left(\sqrt{2-\sqrt{2}}\right)^n$$

Finally note that $e_1 = e_2 = e_3 = 0$ and $e_4 = 2$. From $e_1 = 0 = e_3 = 0$ we note that $A = B$ and $C = D$.

Thus we are left with

$$A(2+\sqrt{2}) + C(2-\sqrt{2}) = 0$$

and

$$A(2+\sqrt{2})^2 + C(2-\sqrt{2})^2 = 1.$$

Multiply the first with $2+\sqrt{2}$ and subtract to obtain

$$C = -\frac{2+\sqrt{2}}{4\sqrt{2}} \text{ and } A = \frac{2-\sqrt{2}}{4\sqrt{2}}.$$

Putting it together we obtain the desired result. $\square$

Example 78. Let $n \geq 1$ be an integer. A set $S \subset \{0, 1, \ldots, 4n-1\}$ is called *rare*, if, for any $k \in \{0, 1, \ldots, n-1\}$ the following two conditions hold :
  (i) The set $S \cap \{4k-2, 4k-1, 4k, 4k+1, 4k+2\}$ has at most two elements;
  (ii) The set $S \cap \{4k+1, 4k+2, 4k+3\}$ has at most one element.
Prove that the set $\{0, 1, \ldots, 4n-1\}$ has exactly $8 \cdot 7^{n-1}$ *rare* subsets.
(American Mathematical Monthly E3328, Romania TST 2006)

Solution. Let a_n the number of *rare* subsets of $\{0, 1, \ldots, 4n-1\}$ and b_n the number of *rare* subsets of $\{0, 1, \ldots, 4n-1\}$ not containing $4n-1$ or $4n-2$. Let's try to obtain recursion relations.

We notice that the difference $a_n - b_n$ is the number of *rare* subsets that contain just one of $4n-1$ or $4n-2$ because of the second condition applied for $k = n-1$.

Now if the set contains $4n-1$ then we must have from the first condition applied for $k = n-1$ that $S \cap \{4n-6, 4n-5, 4n-4\}$ has at most two elements, since the second condition also forces $4n-3$ to not be in our set. We can either include or not include $4n-4$ in our set, independently of our other choices, so we get $2a_{n-1}$ *rare* subsets of this type.

If the set contains $4n-2$, then again from the first condition we know that $S \cap \{4n-6, 4n-5, 4n-4\}$ has at most one element. If $4n-4$ is in our set then it must not contain $4n-6 = 4(n-1)-2$ or $4n-5 = 4(n-1)-1$ so we end up with b_{n-1} such sets. If $4n-4$ does not belong to the set we just get a *rare* subset of $\{1, 2, \ldots, 4(n-1)-1\}$, and so a total of a_{n-1}.

This gives us $b_{n-1} + a_{n-1}$ sets.

Putting it altogether we obtain

$$3a_{n-1} + b_{n-1} = a_n - b_n. \tag{1}$$

Next we need to get a recursion for b_n. These are *rare* subsets which do not contain $4n-1$ or $4n-2$. The first condition then tells us that $S \cap \{4n-6, 4n-5, 4n-4, 4n-3\}$ has at most two elements. If just one of $4n-4$ or $4n-3$ is in our set or none of them then by just deleting again we end up with *rare* subsets of $\{1, 2, \ldots, 4(n-1)-1\}$, and so there are $3a_{n-1}$. If both of $4n-4$ and $4n-3$ then by deleting we obtain a *rare* subset of $\{1, 2, \ldots, 4(n-1)-1\}$ which cannot contain $4(n-1)-2$ or $4(n-1)-1$, thus we get b_{n-1} such sets.

We obtain

$$b_n = 3a_{n-1} + b_{n-1}. \tag{2}$$

Comparing relations (1) and (2), we see that $b_n = 3a_{n-1} + b_{n-1} = a_n - b_n$. Hence $a_n = 2b_n$. Plugging this back into relation (2), it becomes $a_n = 7a_{n-1}$. Since $a_1 = 8$, we see that $a_n = 8 \cdot 7^{n-1}$.

An alternative solution can be obtained by doing induction on the stronger statement $P(n)$: The number of rare subsets of $\{1, 2, \ldots, n\}$ is exactly $8 \cdot 7^{n-1}$ and exactly half of them do not contain either $4n-1$ or $4n-2$. Using this the recurrence for a_n just becomes $\frac{1}{2}a_{n-1} \cdot 6 + \frac{1}{2}a_{n-1} \cdot 8$, where the first term comes from sets where we have either both or one of $4n-6$ or $4n-5$, and the second comes when we have neither of $4n-6$ and $4n-5$. $\square$

Example 79. Prove that there are more than 8^n n-digit numbers not containing any sequence of digits (of any length) twice in a row.

(MOSP 2001, from Kvant)

Solution. Let us call numbers with this property *good*. Let a_n be the number of *good* n-digit numbers. We will show is that $a_n \geq 8a_{n-1}$ and since $a_1 = 9$, we will obtain that $a_n \geq 8^{n-1}a_1 = 9 \cdot 8^{n-1} > 8^n$ and we will be done.

The key is, like in some of the previous problems, to see what happens when we try to complete a *good* string of $n-1$ digits. We can add any digit at the end, unless we create a repeated string.

Obviously, the repeated string can happen only at the end of the number, since we started with a *good* string of length $n-1$. Now obviously, the length of repeating string can be any number between 1 and $\left[\frac{n}{2}\right]$. Finally, we note that if we trim this repeating string at the end we obtaining a *good* string of length $n-r$. Thus we obtain that the number of ways to complete the *good* string of length $n-1$ with a repeat at the end is equal to $\sum_{k=1}^{\left[\frac{n}{2}\right]} a_{n-k}$.

Thus $a_n = 10a_{n-1} - \sum_{k=1}^{\left[\frac{n}{2}\right]} a_{n-k}$. To finish we can use strong induction to show the desired inequality.

Since $a_2 = 81 \geq 8 \cdot 9 = 8a_1$ the first step is done.

Now, using the inductive hypothesis, we have that $a_{n-j} \leq \frac{1}{8^{j-1}} \cdot a_{n-1}$ for each $j = 1, \ldots, n-1$. Thus, using the recurrence relation, we get

$$a_n \geq a_{n-1}\left(10 - \sum_{j=0}^{\left[\frac{n}{2}\right]-1} \frac{1}{8^j}\right) > a_{n-1}\left(10 - \sum_{j=0}^{\infty} \frac{1}{8^j}\right)$$

$$= a_{n-1}\left(10 - \frac{8}{7}\right) > 8a_{n-1}$$

and the inequality is shown, completing the induction. □

Example 80. A self-avoiding rook walk on a chessboard (a rectangular grid of unit squares) is a path traced by a sequence of moves parallel to an edge of the board from one unit square to another, such that each begins where the previous move ended and such that no move ever crosses a square that has previously been crossed, i.e., the rook's path is non-self-intersecting.

Let $R(m,n)$ be the number of self-avoiding rook walks on an $m \times n$ (m rows, n columns) chessboard which begin at the lower-left corner and end at the upper-left corner. For example, $R(m,1) = 1$ for all natural numbers m; $R(2,2) = 2$; $R(3,2) = 4$; $R(3,3) = 11$. Find a formula for $R(3,n)$ for each natural number n.

(Canada 2008)

Solution. Denote by r_n the sought number $R(3,n)$. The initial values are $r_1 = 1$ and $r_2 = 4$. Assume now that $n \geq 3$. We are looking for a recurrence relation.

We note that a self-avoiding rook walk on $3 \times n$ chessboard falls into one of the following categories :
- One walk upwards $(1,1) \to (2,1) \to (3,1)$.

- Walks that do not enter the cell $(2,1)$; any of these walks must start $(1,1) \to (1,2)$ and finish with $(3,2) \to (3,1)$. Thus deleting the first column we have a bijection with self-avoiding rook walks on a $3 \times (n-1)$. We have r_{n-1} such walks.
- Walks that start with $(1,1) \to (2,1) \to (2,2)$ and never come back to the first row. These have to enter the third row from some $(2,k)$ with $2 \leq k \leq n$ and the move leftwards to $(3,1)$; there are $n-1$ such walks.
- Walks that begin $(1,1) \to (2,1) \to (2,2) \to \ldots (2,k) \to (1,k) \to (1,k+1)$ and then they end $(3,k+1) \to (3,k-1) \to \ldots \to (3,1)$ for some $2 \leq k \leq n-1$. After deleting the first k columns, we note that they correspond to walks on a $3 \times (n-k)$ board. So summing up we get $r_{n-2} + r_{n-3} + \ldots + r_1$ such walks.
- Walks which are horizontal reflections of those of the third type; they begin $(1,1) \to (2,1)$ and never enter the third row until the final cell. Again arguing similarly we obtain $n-1$ such walks.
- Finally, walks that are horizontal reflections of the fourth type; we have again $r_{n-2} + r_{n-3} + \ldots + r_1$ such walks.

Summing up everything, we obtain

$$r_n = 1 + r_{n-1} + 2(n-1) + 2(r_{n-2} + \ldots + r_1).$$

This recurrence relation is tricky since the number of terms on the right side grows as n increases. The trick is to rewrite this equation for $n+1$ and then plug in the equation for n. We have

$$\begin{aligned} r_{n+1} &= 1 + r_n + 2n + 2(r_{n-1} + \ldots + r_1) \\ &= r_n + 2 + r_{n-1} + 1 + r_{n-1} + 2(n-1) + 2(r_{n-2} + \ldots + r_1) \\ &= 2r_n + r_{n-1} + 2. \end{aligned}$$

This type of recurrence relations is not homogeneous, and we have to look for a constant c which would make it homogeneous; namely setting $x_n = r_n + c$ we want to have $x_{n+1} = 2x_n + x_{n-1}$.

Thus $-c = -2c - c + 2$, or $c = 1$. Now, we have to solve the recurrence $x_{n+1} = 2x_n + x_{n-1}$ with $x_1 = 2$ and $x_2 = 5$. The characteristic polynomial is $r^2 - 2r - 1 = 0$, with roots $r_1 = 1 + \sqrt{2}$ and $r_2 = 1 - \sqrt{2}$. Thus

$$x_n = C_1(1 + \sqrt{2})^n + C_2(1 - \sqrt{2})^n.$$

Solving for C_1 and C_2 with the initial conditions, we obtain:

$$r_n = \frac{1}{2\sqrt{2}}(1 + \sqrt{2})^{n+1} - \frac{1}{2\sqrt{2}}(1 - \sqrt{2})^{n+1} - 1. \qquad \square$$

Chapter 10

Graph Theory

Graph Theory, though linked with Combinatorics, is also a fascinating branch of mathematics unto itself. Entire textbooks and courses are dedicated to Graph Theory alone; needless to say, we will only be able to touch on a few select topics in this book. In particular, we introduce some basic terminology and facts about graphs, look at the class of graphs called "trees," and touch on the concept of graph coloring. In addition to building up new skills, throughout our discussion we will be able to apply techniques from earlier chapters such as basic counting principles, induction, and the Pigeonhole Principle.

Basic Terminology

Formally, a *graph* $G = (V, E)$ is a pair such that V is a finite set of *vertices* (we represent these as points) and E is a set of *edges* (i.e., 2-element subsets of V). We denote an edge between vertices u and v in G by uv or vu. For our purposes, we focus on "simple graphs": graphs which do not contain loops (edges from a vertex to itself) or multiedges (multiple edges between two particular vertices) and whose edges do not have a direction (i.e., if u and v are vertices of G, then uv and vu are the same edge).

Visually, we use points to represent vertices and represent edges by drawing a line connecting the corresponding two vertices. There are many different ways to draw a graph, but as long as the vertex and edge sets are the same, it is the same graph. For example, below are two different ways to draw the same graph.

Both graphs above are K_4, the complete graph on 4 vertices. In general, the *complete graph* K_n on n vertices is the graph on n vertices with all possible edges.

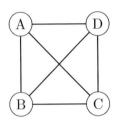 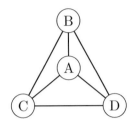

We say two vertices $u, v \in V$ are *adjacent* if there is an edge between them (i.e. $uv \in E$). The *degree* of a vertex $v \in V$ (denoted $d(v)$) is the number of edges of which v is an endpoint.

Example 81 (Handshaking Lemma)**.** The sum of the degrees of all vertices in G is equal to two times the number of edges in G. That is

$$\sum_{v \in V(G)} d(v) = 2|E|.$$

Proof. We prove this fact by counting pairs of the form (v, e) where v is a vertex of G and e is an edge incident to v.

Answer 1. First, we count by focusing on each vertex. How many pairs will contain a particular vertex v? v will appear exactly once for each incident edges; thus v appears in exactly $d(v)$ pairs. Summing over all possible v gives us $\sum_{v \in V(G)} d(v)$ pairs.

Answer 2. Now we count by focusing on each edge. A particular edge is incident upon exactly two vertices, so each edge e appears in exactly two pairs. Thus there are a total of $2|E|$ pairs.

Since these two expressions count the same thing, they must be equal. Thus we have $\sum_{v \in V(G)} d(v) = 2|E|$ as desired. □

Example 82. How many edges are there in the complete graph K_n? What is the degree of each vertex of K_n?

Solution. The complete graph K_n is the graph on n vertices containing all possible edges. This means there exists exactly one edge for every pair of vertices, so a K_n has precisely $\binom{n}{2}$ edges.

Alternatively, since every vertex in the graph will be adjacent to every other vertex in a K_n, the degree of each vertex is $n-1$. Thus the Handshaking Lemma tells us that

$$|E| = \frac{1}{2} \sum_{v \in K_n} d(v) = \frac{1}{2} \sum_{v \in K_n} (n-1) = \frac{n(n-1)}{2} = \binom{n}{2}$$

which agrees with our previous solution. □

A *subgraph* H of a graph G is a graph such that $V(H) \subseteq V(G)$ and $E(H) \subseteq E(G)$.

Example 83. Consider K_6, the complete graph on 6 vertices, with each of the $\binom{6}{2} = 15$ edges colored either red or blue. Show that there exists a monochromatic K_3 as a subgraph regardless of how we color the edges. Show that this is not necessarily true for a K_5.

Solution. Consider a specific vertex v. Since our graph is a K_6, we know that every vertex has degree 5, so by the Pigeonhole Principle v is an endpoint of at least 3 edges that are the same color. Say without loss of generality that there are at least 3 red edges; call the vertices adjacent to v by these edges u_1, u_2, and u_3.

If any edge between the u_i is red, those u_i and v are the vertices of a red K_3. Otherwise, $u_1 u_2, u_1 u_3$, and $u_2 u_3$ are all blue, so u_1, u_2, and u_3 are the vertices of a blue K_3. Thus, no matter how we color the edges, K_6 must contain either a red K_3 or blue K_3. This concludes our proof.

As a final step we show that there exists a coloring of a K_5 with no red K_3 or blue K_3. If we draw K_5 in the "usual" way (with the vertices forming the vertices of a regular pentagon), we can color the outer pentagon one color and the inner star the other color.

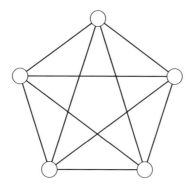

□

A *walk* in a graph is a sequence of adjacent vertices. A *path* is a walk such that all vertices are distinct; a *cycle* is a walk such that all vertices are distinct except that the last vertex is the same as the first vertex. We say a graph G is *connected* if for all pairs of vertices $(u, v) \in V(G)$, there exists a path from u to v. A *component* of G is a maximal (i.e., we cannot add anything to it) connected subgraph of G.

Trees

A graph is called a *tree* if it is connected and acyclic (i.e. it contains no cycles). A degree 1 vertex in a tree is called a *leaf*. Trees are a particularly useful type of graph; they are applied commonly in computer science (both as data structures and as the subject of algortihms). Let's examine some interesting properties of trees.

Example 84. Let T be a tree with at least 2 vertices. Prove that T must have at least 2 leaves.

Solution. In this proof, we focus on extremes and derive a contradiction. Let $P = (v_1, v_2, \ldots, v_{k-1}, v_k)$ be a longest path in T. We claim v_1 and v_k are leaves.

Suppose by way of contradiction that v_1 is not a leaf. Then v_1 has a neighbor in T other than v_2 (call it w). Since T is acyclic, $w \notin P$. But this means $(w, v_1, \ldots, v_k)$ is a longer path than P, contradicting the fact that P was a longest path in T. Thus it must be the case that our assumption was flawed, and v_1 is in fact a leaf. By similar logic, we conclude that v_k must be a leaf. $\square$

Example 85. Show that if T is a tree with a leaf ℓ then the graph $T - \ell$ (the graph formed by removing ℓ and the edge it is an endpoint of from T) is also a tree.

Solution. In order to prove $T - \ell$ is a tree, we must show that it is connected and acyclic. Note that if there were a cycle in $T - \ell$, this cycle would also exist in T when we added back ℓ and its incident edge. Thus $T - \ell$ is acyclic. Now consider any two vertices $u, v \in V(T - \ell)$. We know there exists some (u, v)-path in T; this path will still exist in $T - \ell$ unless one of the vertices in the path is ℓ.

Since ℓ is a leaf, we know it has exactly one neighbor in T, say w. Then any (u, v)-walk passing through ℓ must have the form $(u, \ldots, w, \ell, w, \ldots, v)$. However, since the vertex w appears twice in our sequence, this is not a path at all! Thus we conclude that there must exist some (u, v)-path that does not pass through ℓ, and thus this path exists in $T - \ell$. Since this holds for any $u, v \in V(T - \ell)$, we can conclude $T - \ell$ is connected.

Since $T - \ell$ is connected and acyclic, it must be a tree as desired. $\square$

Example 86. Prove that a tree on n vertices has exactly $n - 1$ edges.

Solution. We proceed by induction on n.
<u>Base Case:</u> ($n = 1$) The single vertex tree has no edges, so our base case holds.
<u>Induction Hypothesis:</u> Suppose that for some $k \geq 1$, all trees on $n = k$ vertices have exactly $k - 1$ edges.

Inductive Step: Consider a tree T on $n = k + 1$ vertices. Since $k + 1 \geq 2$, we know by Example 83 that T has a leaf ℓ. By Example 84 we know $T - \ell$ is a tree on k vertices, so by our induction hypothesis, $T - \ell$ has exactly $k - 1$ edges. T has exactly one more edge than $T - \ell$, namely the edge which has one endpoint that is ℓ. Thus T must have k edges, as desired.

By the principle of mathematical induction, this concludes our proof. $\square$

Graph Coloring

For $k \geq 1$ we say a graph $G = (V, E)$ has a *proper k-coloring* if there exists a mapping $f : V \to \{1, 2, \ldots, k\}$ such that $uv \in E$ implies $f(u) \neq f(v)$ (i.e., we can assign each vertex one of k colors such that adjacent vertices are different colors). We say G is *k-colorable* if G has a proper k-coloring. The *chromatic number* of G (denoted $\chi(G)$) is the minimum k such that G has a proper k-coloring.

Example 87. Determine $\chi(K_n)$ and $\chi(\overline{K_n})$ where $\overline{K_n}$ is the graph on n vertices with no edges.

Solution. Since $\overline{K_n}$ contains no edges, we can simply color all vertices the same color and no two adjacent vertices will be the same color. Thus $\chi(\overline{K_n}) = 1$.

Clearly K_n is n-colorable, since with n colors we could simply assign each vertex a different color. Thus $\chi(K_n) \leq n$. Suppose we attempt to color K_n with fewer than n colors. By the Pigeonhole Principle, we would have to have two vertices of the same color. However, since every pair of vertices in K_n are adjacent, this would not be a proper coloring. Thus n is the minimum number of colors we can use to get a proper coloring, so $\chi(K_n) = n$. $\square$

For all positive integers k, the *chromatic polynomial* of a graph $G = (V, E)$ (denoted $\chi(G; k)$) is the function that counts the number of distinct colorings of G using k or fewer colors. For all $k < \chi(G)$, we know $\chi(G; k) = 0$.

Example 88. Find a formula for $\chi(P_n; k)$ where P_n is the path on n vertices.

Solution. Suppose the vertices of P_n are $v_1, v_2, \ldots, v_n$ such that v_i and v_{i+1} are adjacent for $1 \leq i \leq n - 1$. We color these vertices in order. There are k options for the color assigned to v_1. v_2 only has $k - 1$ options since it may not be the same color as v_1. Similarly, for each v_i with $i > 2$ we have $k - 1$ options since the only restriction is that v_i may not be the same color as v_{i-1}. By the Product Rule, we have $\chi(P_n; k) = k(k - 1)^{n-1}$. $\square$

If $G = (V, E)$ is a graph with edge $e = uv \in E$, then $G \cdot e$ (pronounced "G contract e") is the graph obtained from G by replacing u and v with a single vertex w. w is adjacent to all vertices that are adjacent to either u or v (or both) in G. We delete any multiedges or loops.

Example 89. Let e be an edge of graph G. Prove that

$$\chi(G; k) = \chi(G - e; k) - \chi(G \cdot e; k).$$

Solution. We will prove that $\chi(G - e; k) = \chi(G; k) + \chi(G \cdot e; k)$ by counting the number of proper k-colorings of $G - e$ in two different ways.
Answer 1: There are $\chi(G - e; k)$ proper k-colorings of $G - e$ by the definition of the chromatic polynomial.
Answer 2: Suppose the endpoints of e are u and v. We look at two cases: the colorings where u and v are different colors and the colorings where u and v are assigned the same color. If u and v are assigned different colors, our coloring of $G - e$ must also be a proper coloring of G, and we know the number of such colorings is $\chi(G; k)$.

On the other hand, if u and v are assigned the same color, we can essentially treat u and v as a single vertex of that color, so our coloring must be a proper coloring of $G \cdot e$. We know in total there are $\chi(G \cdot e; k)$ such colorings. Thus in total we have $\chi(G; k) + \chi(G \cdot e; k)$ proper k-colorings of $G - e$.

Since these two answers count the same quantity, they must be equal. This proves our claim. $\square$

We can apply the identity from the previous example to help us derive the chromatic polynomial for cycle graphs. Note that this is in fact the same formula from Example 72, though we describe and prove the result differently here.

Example 90. Prove that $\chi(C_n; k) = (k-1)^n + (-1)^n(k-1)$ where C_n is the cycle on n vertices.

Solution. We prove this by induction on n, the number of vertices in the cycle.
Base Case: ($n = 3$) We have k choices for the color of our first vertex in our C_3. The next vertex is adjacent to the first so we have only $k-1$ color choices. The final vertex is adjacent to both the previous vertices leaving us with $k-2$ options. By the Rule of Product, this yields

$$\chi(C_3; k) = k(k-1)(k-2) = k^3 - 3k^2 + 2k = (k-1)^3 + (-1)^3(k-1)$$

so our base case holds.
Induction Hypothesis: Suppose that for some $\ell \geq 3$ our result holds for $n = \ell$, that is $\chi(C_\ell; k) = (k-1)^\ell + (-1)^\ell(k-1)$.
Inductive Step: Consider the graph $C_{\ell+1}$. By Example 88 we know that for an edge e in a graph G,

$$\chi(G; k) = \chi(G - e; k) - \chi(G \cdot e; k).$$

Let G be $C_{\ell+1}$ with $V(C_{\ell+1}) = \{v_1, v_2, \ldots, v_\ell, v_{\ell+1}\}$ and
$$E(C_{\ell+1}) = \{v_i v_{i+1} \text{ for } 1 \leq i \leq \ell\} \cup \{v_{\ell+1} v_1\}.$$

Let $e = v_{\ell+1} v_1$. We now apply the above formula.

First notice that $C_{\ell+1} - e$ is simply the path graph on $\ell + 1$ vertices or $P_{\ell+1}$. We know $\chi(P_{\ell+1}; k) = k(k-1)^\ell$. Also, $C_{\ell+1} \cdot e$ is simply the graph C_ℓ. We know by our Induction Hypothesis that $\chi(C_\ell; k) = (k-1)^\ell + (-1)^\ell (k-1)$. Substituting this in, we have

$$\begin{aligned}\chi(C_{\ell+1}; k) &= \chi(P_{\ell+1}; k) - \chi(C_\ell; k) \\ &= k(k-1)^\ell - [(k-1)^\ell + (-1)^\ell (k-1)] \\ &= (k-1)^{\ell+1} + (-1)^{\ell+1}(k-1)\end{aligned}$$

as desired. By the principle of mathematical induction, we conclude that $\chi(C_n; k) = (k-1)^n + (-1)^n (k-1)$ for all $n \geq 3$ and all k. $\square$

Chapter 11

Invariants

Problems of this sort involve a process of some type. We can try to associate to the problem a number, quantity that remains invariant under the transformations involved in the process. Usually an invariant problem is pretty easy once you find the right invariant (or a monovariant), but finding it can be pretty tough! Actually, finding the right invariant is an art, and that is what makes these problems hard and only through practice you can get a better grip on imagining and constructing invariants. However there are a few categories of invariants we can always look out for:

- Collorings: Color all the squares in a grid with two or more colors. Usually the chessboard pattern is a good choice, but other patterns are also sometimes useful. Consider squares of each color separately.

- Algebraic expressions: Given a set of values, look at their differences, their sum, the sum of their squares, or their product. If you are working with integers, try looking at these values modulo some number n, usually we pick a prime or a prime power.

- Corners and edges: For grid-based problems, consider any shapes formed. How many boundary edges do they have? How many corners?

- Inversions: If you are permuting a sequence of numbers, consider the number of inversions, that is, the number of pairs (i, j) such that i and j are listed in reverse order. Both the absolute number of inversions and its parity are useful.

- Integers and rationals: Can you cook up from the problem a positive integer that keeps decreasing? Or does the denominator of a rational number keep decreasing?

- Symmetries: Can you ensure that after each step, a figure is symmetrical in some way? Maybe you can divide the objects you have in categories according to similar properties. Perhaps the problem can be divided into two essentially identical subproblems? This is especially useful for game-theory type problems.

Let us get through some examples and as we go along the reader will get a better picture of what this type of problems involve.

Example 91. Several copies of 1 and -1 are written on a board. Alice chooses randomly two numbers written on the board, x and y, deletes them and writes instead 1 if the numbers were equal and -1 otherwise. This process is repeated until only one number is left on the board. Prove that the outcome does not depend on the order in which Alice did the operations.

Solution. Let us do some experimenting. Take for example the sequence of number $(1, 1, -1, -1)$. Then a possible sequence of moves is $(1, 1, -1, -1) \to (1, -1, -1) \to (1, 1) \to 1$. Another one is $(1, 1, -1, -1) \to (1, 1, 1) \to (1, 1) \to 1$.

If we look carefully we see that at each stage the product of the number remains invariant. Let us prove indeed that this happens. If we have two equal numbers, either both are 1 or both are -1 so their product is 1, and indeed it stays the same. If we have two distinct numbers then their product is -1, and again replacing them with -1 preserves their product.

Thus indeed no matter what operations Alice does, at the end the number written on the board will be the product of the numbers initially written. $\square$

Example 92. An 8×8 chessboard has two opposite corners removed. Is it possible to tile it using 2×1 dominoes, without overlaps?

Solution. This is a classical example for coloring the chessboard in the standard way, namely alternatively black and white. Initially we have 32 white squares and 32 black squares. Removing two corners means removing two squares of the same color, say black. Thus $W - B = 2$.

Now, we note that any domino we place covers a white and a black square. Thus, if we cover every square, we must have that the numbers of whites and blacks are the same, $W = B$.

This gives a contradiction and we are done. $\square$

Example 93. The numbers $1, 2, \ldots, 2014$ are written on a sheet of paper in order. Bob randomly picks two numbers and swaps them. Is it possible to reach the initial order of the numbers after 2013 moves?

Solution. This is typical problem for counting inversions, that is we are looking at the quantity I of $x > y$ such that x is on a lower position than y. Now any swap at two numbers k positions apart can be obviously realized as a sequence of $2k - 1$ swaps of adjacent numbers.

Anytime Bob makes swaps for adjacent number either I increases by 1, if the numbers were in order according to the positions or decreases by 1 if the swap is done for an inversion.

Now let us go back to our numbers k distances apart. If we denote by a the 1 increments in I and by b the -1 decrements in I, we have that I modifies by $a - b$. Now since $a + b$ is $2k - 1$, the number of consecutive swaps, it follows that the parity of I changes.

Thus after 2013 moves we have that I is an odd number. But there are no inversions if the numbers are in order, so this means we cannot reach the initial order. $\square$

Example 94. Is it possible to tile a 2014×2014 board with **L** shaped tetrominoes

where the pieces can be flipped or turned?

Solution. The key this time is to color the rows alternatively black and white. Notice that we have an equal number of black and white squares namely $2 \cdot 1007^2$.

Now let us what happens when we place an **L** shaped tetromino on the board. If it not rotated by $90°$, it either covers two white rows and one black, thus it has 3 white squares and one black or it covers two black rows and one white, thus it has 3 black squares and one white. On the other hand if indeed it is rotated by $90°$, then it has three cells on the same row, and thus it covers three white cells and one black, or it covers three black cells and one white cell.

We shall call a tetromino black or white, according to the dominating color of the squares in it. Because we have the same number of black and white squares in our board, if we were able to tile it with tetrominoes then we should have also the same number of white and black tetrominoes, thus an even number of tetrominoes.

But this implies that the area of our board should be divisible by 8, which is not the case since it is equal to $4 \cdot 1007^2$. $\square$

Example 95. Two distinct positive integers a and b are written on a board. The smaller of them is erased and instead we write $\dfrac{ab}{|a-b|}$. The process is repeated as long as the numbers written on the board are distinct. Prove that eventually the process terminates.

(Russia 1998)

Solution. We first do some experimenting.

For $(3,5) \to \left(5, \frac{15}{2}\right) \to \left(\frac{15}{2}, 15\right) \to (15,15)$. For $(4,5) \to (5,20) \to (20,20)$. For $(3,6) \to (6,6)$. For $(2,8) \to \left(\frac{8}{3}, 8\right) \to (4,8) \to (8,8)$.

After some computations we can see that the pattern seems to end up always in the least common multiple of the two numbers we've chosen. Let us prove that.

First, note that we can write the numbers on the board as $\left(\dfrac{ab}{x}, \dfrac{ab}{y}\right)$ and let us on a second board to keep track of the operations on the numbers x, y and we always consider $x > y$ (otherwise swap the order). We claim that after any erasure the effect on the second board is that we replace (x, y) by $(x-y, y)$ or $(y, x-y)$, depending on the order.

To see this note that

$$\frac{\dfrac{ab}{x} \cdot \dfrac{ab}{y}}{\left|\dfrac{ab}{x} - \dfrac{ab}{y}\right|} = \frac{ab}{|x-y|}$$

so the claim is proved.

The familiar reader will now recognize that the process for (x, y) is the Euclidean algorithm. Let us prove that this actually returns the greatest common divisor of (x, y). Since we've started with (a, b) this will end the problem.

So let $a < b$. By the remainder theorem $b = ac_1 + b_1$ with $0 \leq b_1 < a$. If $b_1 = 0$, after $c_1 - 1$ steps we end up with (a, a) and we are done. Otherwise after c_1 steps we will have the tuplet (b_1, a) with $b_1 < a$. Now we repeat $a = b_1 c_2 + a_1$ and we obtain the tuplet (a_1, b_1). Now obviously at any step the bigger number in the tuple decreases so this process has to terminate. It is also clear that we will end up with $(m, 0)$ so in the step before the two numbers had to be equal.

Finally, note that the operation leaves the greatest common divisor of the two numbers invariant, so the two numbers we obtain are actually equal to $\gcd(a, b)$ and we are done. □

Example 96. Several candies are placed on an infinite (in both directions) strip of squares. As long there at least four candies on a square, you can pick up four of them and distribute two to the preceding square and two to the following square. Is it possible to return to the original configuration after a finite sequence of moves?

Solution. If you start experimenting with the problem you note at first glance, that either a phenomenon like a huge pile candies appears in one of the squares or either some of candies keep being pushed farther apart from the initial square where they were at the beginning. To prove that we cannot return to the initial configuration, the natural thing is to look for some quantity that increases with every move. Then this quantity cannot return to its initial value, and hence we cannot return to the starting position. Notice that the basic move tends to force candies apart. Thus we need to find a quantity that detects this.

Label the strip with consecutive integers and let c_i the label of the square containing candy number i. Let us consider the quantity $C = \sum c_i^2$. When we are doing a move and remove four candies from the square with label a, we note that C drops by $4a^2$. When we place the candies, we observe that the new C increases by $2(a-1)^2 + 2(a+1)^2 = 4a^2 + 4$ since we are placing two candies to the left and right of our square a.

Thus for a move we have that the quantity C increases by 4 so we can never return to the initial configuration, after a finite sequence of moves. □

Example 97. We have 2014 teddy bears distributed randomly in 127 boxes. Each minute, as long as not all teddy bears are in the same box, we move a teddy bear from a box to a different box with at least as many teddy bears as the box we've picked. Prove that eventually all the teddy bears will be gathered in one box.

Solution. First, let us note that the result of the problem is plausible. The boxes with a lot of teddy bears will tend to get more bears, while the boxes with a few will tend to lose teddy bears. So the outcome should be that all teddy bears end up in one box. The point is how to make this reasoning rigorous.

We need to construct a monovariant like in the previous problem. Since the number of teddy bears remains invariant, we consider the sum $T = \sum_{i=1}^{127} t_i^2$, where t_i is the number of teddy bears in box i. The claim is that T increases with each move.

For this, suppose we remove one teddy bear from a box with n bears and add it to a box with $m \geq n$ teddy bears. The net effect on T is that we add the difference between $(m+1)^2 + (n-1)^2$ and $m^2 + n^2$. Note that $(m+1)^2 + (n-1)^2 - m^2 - n^2 = 2m - 2n + 2 \geq 2$, so T increases by at least 2.

Now, the number of distributions of 2014 teddy bears into 127 boxes is finite, so there are only finitely many possibilities for T. This means that T cannot increase forever. But as long as we have two boxes with teddy bears, we can repeat the process and increase T, thus the process has to terminate with all teddy bears in the same box. □

The next problem also introduces a useful idea similar to coloring, namely assigning weights.

Example 98. There are n markers, each with one side white and the other side black, aligned in a row with their white sides up. At each step, if possible, we choose a marker with the white side up (but not one of the outermost markers), remove it, and reverse the two neighboring markers. Prove that one can reach a configuration with only two markers if and only if $3 \nmid n - 1$.

(IMO Shortlist 2005)

Solution. Denote a state of the markers by a string of the characters W, B, and x, where a W denotes a white side up, a B a black side up, and an x a deleted marker. We include the x label to help us keep track of the number markers removed. Missing markers are ignored, so these x's are not really necessary.

We first experiment with what we can achieve by a sequence of moves. The sequence

$$WWWWWW \cdots \Rightarrow BxBWWW \cdots \Rightarrow BxWxBW \cdots \Rightarrow WxxxWW \cdots$$

shows that if we have $n \geq 5$ white markers, then we can always reach a state with $n - 3$ white markers. Thus if $n = 3m$, we can iterate this to reach a state with 3 white markers, and one additional move will leave us with 2 black markers. If $n = 3m + 2$, we can iterate this to reach a state with 2 white markers.

To complete the problem, we need to show that if $n = 3m + 1$, we cannot reach a state with just two markers. Iterating the construction above will not get us to two markers, but this doesn't prove anything since there are many other sequences of moves we could apply. For this part of the problem we will need an invariant.

The first invariant one probably notices is that the number of black markers is always even. This is easy to prove. A move deletes a white marker (which does not change the parity of the number of blacks) and flips two markers. Any flip of a marker changes the parity of the number of blacks, so two flips combine to give no net change to the parity. This proves that (as we saw above) if we end up with two markers, then they are the same color.

Unfortunately, this invariant alone does not solve the problem. We need a more refined invariant. Assign a weight to every marker as follows. All black markers get assigned weight 0. A white marker gets a weight of $+1$ if there are an even number of black markers to its left and weight -1 if there are an odd number of black markers to its left. Let S be the sum of the weights of all the markers.

Let us look at how the sum S changes if we do a move. Suppose we do a move on a white marker with weight w. The first thing to notice is that if a marker was unchanged by the move, then its weight is also unchanged. This follows from the same parity argument above and we will not repeat it. Thus we need only analyze the three markers involved in the move. The effect on the central marker is easy: it goes from contributing w to being gone and hence contributing 0. For the left marker, there are two cases. If the left marker was white before the move, then it also had weight w before the move and after the move it is black and hence has weight 0. If the left marker was black, then it had weight 0 before the move and after the move it has weight $-w$ (since it has one fewer black marker to its left than the central marker had). In either case we see that the contribution of the left marker decreases by w. For the right marker the cases are similar and we also get a net decrease of w. If the right marker was originally white, its weight went from w to 0. If it was originally black, its weight went from 0 to $-w$ (since the flip of the left marker means that the parity of the right marker after the move will be opposite to the central marker's parity before the move). Thus overall we see that S decreases by $3w$. Hence S remains unchanged modulo 3.

This is the invariant we needed. If $n = 3m + 1$, then we started with all white markers of weight 1 and hence $S = 3m + 1 \equiv 1 \pmod{3}$. However, if we end with two markers, then we have either two white markers $S = 2$ or two black markers $S = 0$. Thus we cannot reach a configuration with two markers. □

Example 99. A $n \times n$ grid is given, $n1$ squares of which contain a one, and the rest of the squares contain a zero. It is allowed to select a square, subtract 1 from the number in that square, and add 1 to all numbers in the same row and column as this square. Is it possible to get a grid where all numbers in the squares are equal?

(Russia 1998)

Solution. No. We first note that there must be a 2×2 square in the original grid with three zeroes and one 1. This follows from the fact that there is a row containing only zeroes, hence there is a row containing only zeroes adjacent to a row containing a 1. Call this square special.

Now, consider any 2×2 square in the grid, call it K. Let a, b, c, d be the numbers in it, starting from left top corner to lower right corner going along the rows. Consider $S = (a+d) - (b+c)$. We shall prove that S stays invariant modulo 3.

Now, let's see what happens when an operation occurs. If the square we decrease is not a part of K, it is easy to see that $S = S'$, since it will either modify no entry in K or just modify two values by incrementing them with 1 and they will be on opposite sides in S.

When the square we decrement is a part of S, we can assume it's in the top left corner, since the situation it's symmetrical. Then $a' = a - 1$, $b' = b + 1$, $c' = c + 1$ and $d = d'$. Thus we see that $S' = S - 3$ and the claim is proved.

Now we see that in the special square we can never make the numbers equal, since S should remain equal to 1 modulo 3 and we are done. □

Example 100. A regular (5×5)-array of lights is defective, so that toggling the switch for one light causes each adjacent light in the same row and in the same column as well as the light itself to change state, from on to off, or from off to on. Initially all the lights are switched off. After a certain number of toggles, exactly one light is switched on. Find all the possible positions of this light.

(APMO 2007)

Solution. This is another typical invariant problem. This time we have to figure a labeling that will stay invariant and will restrict the possibilities for the light switched on.

Considering the following labeling

1	1	0	1	1
0	0	0	0	0
1	1	0	1	1
0	0	0	0	0
1	1	0	1	1

The invariant we are seeking is parity the sum of the labels of the lights that are turned on. The key of this observation is that every 1 has precisely one neighbor whose label is 1 and every zero has an even number of neighbors labeled 1.

Now we can define a new labeling by reflecting across the diagonal or rotating by 90°

1	0	1	0	1
1	0	1	0	1
0	0	0	0	0
1	0	1	0	1
1	0	1	0	1

A similar argument shows that the parity of the sum of the labels where the light are turned on is an invariant.

Since the initial sum was zero, we need to identify the common zeroes in both of the labelings and these are clearly in the cells $(2,2)$, $(2,4)$, $(3,3)$, $(4,2)$, $(4,4)$.

The only thing left to do is to construct a sequence of toggles that will reach the desired state in these cells. It suffices to do it for the center and $(2,4)$, the rest follow by symmetry.

For $(3,3)$ we should do the moves

			t	t
		t		
	t	t		t
t				t
t		t	t	

Finally, for $(2,4)$, we should do the sequence of moves

	t		t	
t	t		t	t
	t			
		t	t	t
			t	

$\square$

Example 101. To each vertex of a regular pentagon an integer is assigned so that the sum of all five numbers is positive. If three consecutive vertices are assigned the numbers x, y, z respectively, and $y < 0$, then the following operation is allowed: x, y, z are replaced by $x + y$, $-y$, $z + y$ respectively. Such an operation is performed repeatedly as long as at least one of the five numbers is negative. Determine whether this procedure necessarily comes to an end after a finite number of steps.

(IMO 1986)

Solution. Considering a few examples with small numbers, one should quickly guess that the procedure necessarily comes to an end after a finite number of steps. To prove this we look for a monovariant. There are many possible choices for this.

Since the sum stays the same, we would be tempted to try the sum of the squares of the numbers assigned to the vertices. A quick test on examples shows that this will not behave as we wanted. Another wrong guess would be to take the minimum of the numbers. If all x, y, z are negative, the minimum decreases and thus is not a monovariant.

Thus we see that a monovariant should take into account the neighbors of each vertex. Thus we consider

$$I(a_1, a_2, a_3, a_4, a_5) = \sum_{i=1}^{5} a_i^2 + \sum_{i=1}^{5} (a_i + a_{i+1})^2.$$

Assume we have done the operation centered at $a_2 < 0$. Then the new value of I is

$$I(a_1+a_2, -a_2, a_3+a_2, a_4, a_5) = I(a_1, a_2, a_3, a_4, a_5) + 2a_2(a_1+a_2+a_3+a_4+a_5).$$

The value of I is obviously a positive integer, and it decreases with each move since $a_2 < 0$ and we know that the sum of the values at the vertices is positive. Thus there can be only finitely many steps. $\square$

Chapter 12

Combinatorial Geometry

Problems in this area involve a blend of geometry and combinatorics. Useful ideas for arguments are:

- The extremal element method. Namely, pick the configuration which maximizes or minimizes a certain quantity. One must take care to show this extremum exists, but this is usually a consequence of the fact that there are only finitely many possible configurations

- Projections

- Convexity

- Pigenhole principle

Also as we go along, we will prove some classical results; these will have proof instead of solution and they can be as results in any problem or mathematical contest.

Example 102. (Sylvester-Gallai) Consider finitely many points in the plane such that any line passing through two of the given points contains one more of the given points. Prove that all points are on a line.

Proof. We will assume the contrary. Consider pairs of type (P, l) where l is a line which passes through two of the given points, and P is a point in our set with $P \notin l$.

Such pairs exist since the points are not all on a line. There are finitely many such pairs, so we can choose a pair such that $\text{dist}(P, l)$ is minimized. Let C be the foot of the perpendicular from P to l. Note that the line l must contain at least three points from our set, say A, B, and D. We can assume without loss of generality that D and A are on the same side of C with respect to l and that A is between C and D. Then it is easy to see that

$\text{dist}(A, PD) < PC = \text{dist}(P, l)$ and this contradicts our initial choice of the pair (P, l). □

It is often convenient to rephrase the Sylvester-Gallai theorem in a more positive way: If there is a set of n points in the plane, not all collinear, then there is a line which passes through exactly two of them.

Example 103. Prove that $n \geq 3$ points, not all of them collinear, determine at least n lines.

Proof. We will proceed by induction on n. For $n = 3$ it is obvious. Assume the statement is true for n. By the consequence, for the points $\{P_1, \ldots, P_{n+1}\}$, there is a line passing through only two of them, without loss of generality say the line connecting P_{n+1} and P_n. Note that if all of $P_1, \ldots, P_n$ are collinear, we can easily get $n+1$ lines from connecting P_{n+1} to each of them and are done.

Otherwise, removing P_{n+1} we get a set of n points that are not all collinear. Thus we can apply the induction step, and get at least n lines. When we add P_{n+1} back in, we get at least one new line, namely, the line joining P_n and P_{n+1}. Hence we have at least $n+1$ lines. □

Example 104. Consider finitely many lines on the plane with no two parallel such that for the intersection point of any 2 of them at least one more of the given lines passes through it as well. Then all the lines are concurrent.

Solution. Assume the contrary, that not all of the lines are concurrent. Consider pairs (P, l) where P is an intersection point, l is one of the lines in our set, and $P \notin l$.

There are finitely many such pairs, so we can choose a pair which minimizes $\text{dist}(P, l)$. Let C be the foot of the perpendicular from P to l. At least three lines from our set, say l_1, l_2, and l_3, pass through P. Let them cut l at A, B and D, respectively. Without loss of generality, we can assume A and D are on the same side of C with respect to l and that A is between C and D. Then we obtain a contradiction by noting the same inequality as in Example 103, $\text{dist}(A, PD) < PC = \text{dist}(P, l)$. □

Example 105. Prove that any convex polygon $\mathcal{P}$ with area 1 can be included in a rectangle with area 2.

Solution. Let A and B be vertices of $\mathcal{P}$ such that the distance between them is maximal. We will think of the line AB as being horizontal and use the corresponding terminology. Draw the vertical lines through A and B. We first claim that any vertex of the polygon has to be in the band between these two lines: if a vertex X is outside this band, then it is easy to see that either AX or BX is greater than AB and this contradicts our choice of A and B.

Next, let C be a vertex of $\mathcal{P}$ that is farthest above the line AB and D be a vertex of $\mathcal{P}$ that is farthest below the line AB. (Either C or D may agree with A or B if $\mathcal{P}$ lies entirely on one side of the line AB.) Drawing the horizontal lines through C and D gives a rectangle $MNPQ$ containing $\mathcal{P}$ (see the figure).

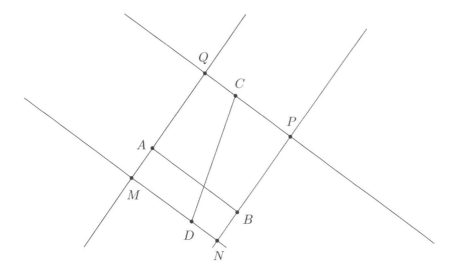

Let $[\cdot]$ denote the area of a figure. The heights of the triangle ABC and the rectangle $ABPQ$ are equal (thinking of AB as the base in both cases). Hence $[ABPQ] = 2[ABC]$. Similarly, $[MNBA] = 2[ADB]$. Also note that A, B, C, and D are among the vertices of $\mathcal{P}$, so $\mathcal{P}$ contains quadrilateral $ADBC$. Hence $[MNPQ] = 2[ADBC] \le 2[\mathcal{P}] = 2$. $\square$

Example 106. Find all finite sets S of points in the plane with the following property: for any three distinct points A, B, and C in S, there is a fourth point D in S such that A, B, C, and D are the vertices of a parallelogram (in some order).

(USA TST 2005)

Solution. We claim that for sets of 3 or more points, only the vertices of a parallelogram satisfy the condition. A set with fewer than 3 points satisfies the condition vacuously.

Assume there is a set S with 3 or more points that satisfies the condition. Note that no three points of S can be collinear, since three collinear points could never be extended to a parallelogram. Take the triangle ABC with vertices among the points of S that has maximal area. For each vertex of ABC draw the parallel to the opposite side through it. These three lines form a triangle DEF with A the midpoint of side EF, B the midpoint of side DF, and C the midpoint of side DE.

The first thing to notice is that the fact that ABC has maximal area implies that S is entirely contained in triangle DEF. To see this, suppose $X \in S$ were outside DEF. Then without loss off generality, X lies on the side of line DE opposite F. Since C is on line DE and AB is parallel to DE, we see that X is farther from line AB than C is. Hence $[ABX] > [ABC]$, contradicting our choice of ABC.

The second thing to notice is that D, E, and F are the three points which with A, B, and C can form the vertices of a parallelogram. Hence by the property of S, one of these lies in S. Assume without loss of generality that it is D. Since $[ABC] = [BCD]$, triangle BCD also has maximal area. Thus we could repeat the argument of the last two paragraphs and find a second triangle containing S. Intersecting this triangle with DEF, we conclude that S is entirely contained in the parallelogram $ABDC$.

Now if S is not just the vertices of a parallelogram, then it must contain a fifth point X. By symmetry, we may assume X lies inside triangle BCD. There must be a Y in S such that B, D, X, and Y (in some order) are vertices of a parallelogram. If the parallelogram is $BXDY$, then the midpoint of XY, which is also the midpoint of BD, lies on line DF. Hence Y lies on the opposite side of line DF from X and hence is outside triangle DEF, contradicting our result above. If the parallelogram is either $BDXY$ or $BDYX$, then XY is parallel to and congruent to BD and AC. Hence Y cannot lie inside parallelogram $ABDC$, a contradiction. □

Another important tool is the use of projections, as we can see from the following example.

Example 107. Inside a unit square several circles are placed, having the sum of their circumferences equal to 10. Prove that there exists a line which intersects at least four of the circles.

Solution. Let O_1, O_2, ..., O_n be the centers of the circles. Consider the projections of each circle on the side AB of the square.

It is easy to see that each projection is given by a diameter of the circle parallel to the side AB. Let P_iQ_i be the segments we have on the side AB. Then by the hypothesis we know that $\sum_{i=1}^{n} l(P_iQ_i) = \dfrac{10}{\pi}$.

If we assume that there do not exist four segments that intersect, this means that each point on the side AB is covered by at most three segments, thus $\sum_{i=1}^{n} l(P_iQ_i) \leq 3AB = 3$. This would mean that $\dfrac{10}{\pi} \leq 3$, and this gives us a contradiction.

Combinatorial Geometry 113

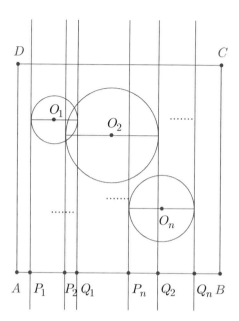

Thus there must be four overlapping segments, and taking the perpendicular line to AB at this intersection point would cut the four circles whose projections are the segments including it. □

Next we move on to another fundamental concept in combinatorial geometry, namely the notion of convex hull for a set of points.

Example 108. If $A_1, \ldots, A_n$, $n \geq 3$, are points in the plane, with no three collinear, then there is a convex polygon P whose vertices are some of the points A_i and the rest of these points are inside P. The polygon P is called the *convex hull* of the points $A_1, \ldots, A_n$.

Proof. We will proceed by induction on n. For $n = 3$, it is obvious since the three points form the vertices of a triangle. Assume the statement is true for sets of n points and consider adding an $(n+1)$-st point A_{n+1}. Let P be the convex hull of the n points. If the $(n+1)$-st point is inside P, we are done, so assume it is outside P. There is some closest point X to A_{n+1} on the boundary of P. (This is not completely obvious, but X must be either a vertex of P or the foot of the perpendicular from A_{n+1} to a side of P. This is only finitely many possibilities, hence one must be closest.) Consider the line ℓ perpendicular to XA_{n+1} at A_{n+1}. The convex polygon P cannot intersect ℓ. If P met ℓ at Y, then the segment XY would be inside P, and somewhere on the perpendicular from A_{n+1} to XY would be a boundary point of P closer to A_{n+1} than X, a contradiction. Hence P lies entirely in one of the half-planes determined by ℓ.

Look at the signed angles that the rays from A_{n+1} through the vertices of P make with the ray $A_{n+1}X$. There will be some vertex B of P that makes the smallest signed angle and some vertex C of P that makes the largest signed angle. Then P lies entirely in the angle $\angle BA_{n+1}C$. The line BC cuts P into two convex polygons (since the cut only decreases interior angles). Let $B_1 = B, B_2, \ldots, B_k = C$ be the vertices of the one that lies on the opposite side of line BC from A_{n+1}. Then the polygon $Q = A_{n+1}B_1B_2 \cdots B_k$ is the desired convex hull. The interior angle at A_{n+1} is less than 180 degrees since B and C are on the same side of ℓ. The angles at B and C are less than 180 degrees, because B_2 and B_{k-1} are inside $\angle BA_{n+1}C$. At all the other vertices the interior angles are unchanged, so they are less than 180 degrees by the induction hypothesis. The interior of Q contains the interior of triangle $A_{n+1}BC$, which in turn contains the part of P on the same side of BC as A_{n+1}. Hence Q contains P and all the points $A_1, \ldots, A_n$ which are not on the boundary of Q are in its interior.

One can also argue fairly easily that the convex hull is unique. A point A_k is a vertex of the convex hull P if and only if there is a line through A_k such that all the other vertices $A_1, \ldots, A_{k-1}, A_{k+1}, \ldots, A_n$ lie on the same side of this line. □

Example 109. There are 5 points in the plane with no three collinear. Prove that you can pick 4 points which are vertices of a convex quadrilateral.

Solution. Consider the convex hull of the 5 points. We have two cases. If the convex hull is a pentagon or convex quadrilateral, then we are done. Thus we may assume the convex hull is a triangle. Let this triangle be ABC and let D and E be the points in its interior. Two of the vertices of ABC are on the same side of the line DE, without loss B and C. Since segment DE lies inside ABC, it cannot be crossed by line BC, hence D and E are on the same side of BC. Further assume without loss that D is closer to side AC than E is. Then $BCDE$ is of a convex quadrilateral. □

Example 110. Find all integers $n > 3$ for which there exist n distinct points $A_1, \ldots, A_n$ in the plane, no three collinear, and real numbers $r_1, \ldots, r_n$ such that for distinct i, j, k the area of the triangle $A_iA_jA_k$ is $r_i + r_j + r_k$.

(IMO 1995)

Solution. Note that if the statement is false for some k points, then it is false any number of points greater than k. It is easy to show that $n = 3$ and $n = 4$ work. For $n = 4$, we have two pictures. If $A_1A_2A_3A_4$ is a convex quadrilateral then expressing the area by splitting along the diagonals we have

$$r_1 + r_2 + r_3 + r_1 + r_4 + r_3 = r_2 + r_1 + r_4 + r_2 + r_3 + r_4$$

or
$$r_1 + r_3 = r_2 + r_4.$$

Otherwise suppose A_4 is inside the triangle $A_1A_2A_3$. Then expressing the area of $A_1A_2A_3$ in two ways, either dividing using A_4 or use the hypothesis we get

$$r_4 = -\frac{r_1 + r_2 + r_3}{3}.$$

Now consider $n = 5$. We analyze the possibilities for the convex hull

- If it is a triangle, without loss of generality say $A_1A_2A_3$, then the two points A_4 and A_5 are inside and by the above

$$r_4 = r_5 = -\frac{r_1 + r_2 + r_3}{3}.$$

But now A_5 is going to lie in one of the triangles $A_1A_2A_4$, $A_1A_3A_4$ or $A_2A_3A_4$. Using the side of the triangle and these two points we see that $r_5 < r_4$, unless the points coincide, which is a contradiction either way.

- If it is a quadrilateral say $A_1A_2A_3A_4$, then looking at the 4 triangles formed by the intersection of the diagonals, we know A_5 has to lie in two of them. Thus A_5 must be in two triangles subtended by the same side of the quadrilateral and assume they are $A_1A_2A_3$ and $A_1A_2A_4$. Then again using the partition of these triangles we have

$$r_5 = -\frac{r_1 + r_2 + r_3}{3} = -\frac{r_1 + r_2 + r_4}{3}$$

so $r_3 = r_4$.

Using $r_1 + r_3 = r_2 + r_4$ we obtain $r_2 = r_1$. Finally we can partition the quadrilateral using A_5 so

$$4r_5 + 2(r_1 + r_2 + r_3 + r_4) = r_1 + r_3 + 2(r_2 + r_4).$$

Thus $4r_5 + r_1 + r_3 = 0$ and looking at $3r_5 + 2r_1 + r_3 = 0$, we see that $r_5 = r_1$.

Then we see that $(A_3A_4A_1) = (A_3A_4A_2) = (A_3A_4A_5)$, where (X) denotes the area of figure X. It is an elementary fact that if they are on the same side of A_3A_4 it follows that $A_1A_5 \| A_3A_4$ and $A_3A_4 \| A_2A_5$, and we get the sought contradiction, we cannot draw two parallels from A_5 to A_3A_4 unless the point are collinear, but this is forbidden by the hypothesis.

- If we have a convex pentagon, $A_1A_2A_3A_4A_5$, then using a vertex and the diagonals emerging from that vertex we can express the area of the pentagon in five different ways; namely

$$3r_1 + 2(r_3 + r_4) + r_2 + r_5, \ 3r_2 + 2(r_4 + r_5) + r_1 + r_3,$$

$$3r_3 + 2(r_1 + r_5) + r_2 + r_4, \ 3r_4 + 2(r_1 + r_2) + r_3 + r_5$$

and

$$3r_5 + 2(r_2 + r_3) + r_1 + r_4.$$

All of these have to be equal to each other and doing the algebra, we see that this forces $r_1 = r_2 = r_3 = r_4 = r_5$.

Then again as above we get a contradiction,

$$(A_1A_2A_3) = (A_1A_2A_4) = (A_1A_2A_5).$$
□

Another useful result on convex sets is Helly's theorem.

Definition. A figure F is convex if for any two points A, B in F, the segment AB is also contained in F.

Example 111. (Helly's theorem) Suppose we have n convex figures with $n > 4$ in the plane, call them $F_1, ..., F_n$ such that $F_i \cap F_j \cap F_k \neq \emptyset$ for any $1 \le i < j < k \le n$. Then

$$\bigcap_{i=1}^{n} F_i \neq \emptyset$$

Proof. We proceed by induction on n. The base case is when $n = 4$. Let $A_4 \in F_1 \cap F_2 \cap F_3$, $A_3 \in F_1 \cap F_2 \cap F_4$, $A_2 \in F_1 \cap F_3 \cap F_4$ and $A_1 \in F_2 \cap F_3 \cap F_4$, be four points in the intersections. We have a few cases to analyze:

- If the four points are collinear. Then we can assume without loss of generality that the order on the line is A_1, A_2, A_3, A_4. Then note that A_1 and A_3 are points in F_2 and F_2 is convex so $A_1A_3 \in F_2$. Now A_2 is in the segment A_1A_3 so $A_2 \in F_2$. Thus A_2 is in the intersection of all the figures.

- If their convex hull is a triangle: Assume that A_4 lies inside the triangle $A_1A_2A_3$. Because the three vertices A_1, A_2, A_3 belong in F_4, it is an easy argument to see that whole triangle, with border and interior is contained in F_4. Thus A_4 is this time in the intersection of all the figures.

- Finally if the convex hull is a convex quadrilateral, without loss of generality $A_1A_2A_3A_4$. Consider the diagonal A_1A_3 and diagonal A_2A_4. Their intersection point P is going to be in the intersection of all the convex figures.

This proves the $n = 4$ case. Assume the statement true for $n > 4$. Then consider $n + 1$ sets such that the intersection of any three is nonempty. Let $J_i = F_i \cap F_{n+1}$ for $1 \leq i \leq n$. The intersection of any three J_i is nonempty because of the $n = 4$ case. Therefore, the intersection of all the J_i is nonempty but this is equal to the intersection of all the F_i's with $1 \leq i \leq n + 1$. □

Let us show an application of this theorem.

Example 112. Given n points in plane such that any three of them can be covered by a unit disk, prove that all n points can be covered by a unit disk.

Solution. Let us denote the points with $P_1, P_2, \ldots, P_n$ and let us construct for each P_i the circle of radius 1 centered at it, call it $\mathcal{C}_i$. Our goal is to prove that $\bigcap_{i=1}^{n} \mathcal{C}_i \neq \emptyset$; if we consider O in the intersection, then drawing the circle of radius 1 centered at it will cover all the points.

Thus since discs are convex figures, by the previous theorem it suffices to prove that any three discs intersect. Assume without loss of generality we are looking at the first three points, since the argument is the same for all pairs of points. We know that there is a circle of radius 1 covering them and call its center S. Then since $P_1S, P_2S, P_3S \leq 1$ it follows that $S \in \mathcal{C}_1 \cap \mathcal{C}_2 \cap \mathcal{C}_3$ and we are done. □

Finally we end with some diverse problems, which contain in their solutions useful and nice ideas.

Example 113. Let $n \geq 3$ be a fixed natural number. Find all functions $f : \mathbb{R}^2 \to \mathbb{R}$ such that for any $A_1, \ldots, A_n$ which are vertices of a regular n-gon, we have
$$f(A_1) + f(A_2) + \ldots + f(A_n) = 0$$

(Romania TST 1996)

Solution. We will show that the only possibility is that $f = 0$. Let A be an arbitrary point. Consider a regular n-gon $AA_1A_2 \ldots A_{n-1}$. Let k be an integer, $0 \leq k \leq n-1$. A rotation with center A of angle $\dfrac{2k\pi}{n}$ sends the polygon $AA_1A_2 \ldots A_{n-1}$ to $A_{k0}A_{k1} \ldots A_{k,n-1}$, where $A_{k0} = A$ and A_{ki} is the image of A_i, for all $i = 1, 2, \ldots, n-1$.

From the condition of the statement, we have
$$\sum_{k=0}^{n-1} \sum_{i=0}^{n-1} f(A_{ki}) = 0.$$

Observe that in the sum the number $f(A)$ appears n times, therefore

$$nf(A) + \sum_{k=0}^{n-1}\sum_{i=1}^{n-1} f(A_{ki}) = 0.$$

On the other hand, we have

$$\sum_{k=0}^{n-1}\sum_{i=1}^{n-1} f(A_{ki}) = \sum_{i=1}^{n-1}\sum_{k=0}^{n-1} f(A_{ki}) = 0,$$

since the polygons $A_{0i}A_{1i}\ldots A_{n-1,i}$ are all regular n-gons (they are all centered at A and obtained by rotating around A the point $A_{0,i}$). From the two equalities above we deduce $f(A) = 0$, hence f is the zero function. □

Example 114. (Erdos-Anning) For any natural number n there exists a set of n points not all of them collinear, such that the distance between any two is an integer. Prove also there is no infinite set with this property.

Solution. To construct such a set of n points, we will take equally spaced points around a circle. This has the advantage of giving fewer distinct distances. We will actually find points on the unit circle with center at the origin with the distances between them all rational numbers. We can then find the desired points by doing a high enough power homothety to clear the denominators.

We know that there exists an angle θ such that $\sin(\theta) = \dfrac{3}{5}$ and $\cos(\theta) = \dfrac{4}{5}$. Now starting from the positive x-axis, we construct the point P_k to have a center angle congruent to $2k\theta$ modulo 2π (for big k this means we have to go around the circle). Now the distance between P_i and P_j is

$$P_iP_j = 2|\sin((i-j)\theta)|.$$

It is easy to see that $\sin(k\theta)$ is rational for any integer k. One way is from de Moivre's formula

$$\cos(k\theta) + i\sin(k\theta) = \left(\frac{4+3i}{5}\right)^k.$$

Another way is to induct using the formula

$$\sin((k+1)\theta) = 2\cos(\theta)\sin(k\theta) - \sin((k-1)\theta).$$

The only subtle point is that in our construction, since we go around the circle, points might overlap. This would make some of our distances equal to zero, so it can only happen if $\sin(k\theta) = 0$ for some k. However, if you take a little more care when proving $\sin(k\theta)$ is rational, you can show more, namely that $\sin(k\theta) = \dfrac{m_k}{5^k}$, where m_k is an integer not divisible by 5. Hence in particular, $\sin(k\theta) \neq 0$.

For the second statement, assume that such an infinite set exists, and pick two points A and B. By the triangle inequality for another point P, we have that $|PA - PB| \leq AB$. By the Pigeonhole Principle, there is a fixed integer k for which there exist infinite many P with $PA - PB = k$. Consider the geometric locus of all such points P, which forms a hyperbola, call it $\mathcal{H}_1$. Now restrict your attention to only the points on $\mathcal{H}_1$. This is again an infinite set. Thus, picking C and D on $\mathcal{H}_1$ and using the same argument, we find a different hyperbola $\mathcal{H}_2$ which contains infinitely many points from $\mathcal{H}_1$. But it is a well-known geometry fact, that distinct hyperbolas intersect in at most four points, and the set of intersections we found was infinite, and this gives us the desired contradiction. $\square$

Example 115. Prove that n points in the plane determine at most $cn^2\sqrt{n}$ right triangles, for some constant $c > 0$.

(Miklos-Schweitzer)

Solution. We shall denote by $f(n)$ the maximal number of right triangles for any configuration of n points in the plane. Our goal is to prove $f(n) \leq n^2\sqrt{n}$. We have that the statement is trivial for $n = 1, 2, 3, 4, 5$. Assuming the contrary and there would be a smallest n such that $f(n) > n^2\sqrt{n}$.

Let us take the points $P_1, P_2, \ldots, P_n$ on the plane that give the number $f(n)$. We claim that there must be a straight line containing $2\sqrt{n}$ of the n points.

Consider all the possible ordered tuples (P_i, P_j) and assign to each the the number of right triangles $P_iP_jP_k$ such that the right angles is in P_i. This number is the number of points on the perpendicular line through P_i to P_iP_j. The sum of these is twice the number of right triangles, namely bigger than $2n^2\sqrt{n}$.

On the other hand since the number of terms in the sum is at most the number $n(n-1)$, one of the sumands has to contain at least $2\sqrt{n}$ terms and we are done.

Now let's eliminate the points on this straight line and analyze how many right triangles disappear. We have two groups:

- Triangles for which the right angled vertex is omitted. We have $\dfrac{n(n-1)}{2}$ choices for a hypotenuse of such a triangle and a segment can serve as a hypotenuse for for at most two right triangles of this type so the number of triangles we cut off is at most $n(n-1) < n^2$.

- Triangles with right angled vertex lying off the line and one of the vertices lying on the line. In this case we have $\dfrac{n(n-1)}{2}$ choices for those vertices that are not required to lie on the straight line, and we can choose at most two points on the critical line such that it forms a triangle of this type together with a fixed couple P_iP_j so that the right angle is at P_i or P_j. Thus also for this subcase we have at most $n(n-1) < n^2$ triangles.

Thus the removal of the points on the straight line destroys at most $2n^2$ right triangles thus the remaining $n - 2\sqrt{n}$ points span at least $n^2\sqrt{n} - 2n^2$ right triangles. But now by the choice of n we must have that

$$n^2\sqrt{n} - 2n^2 \leq (n - 2\sqrt{n})^2\sqrt{n - 2\sqrt{n}}$$

and solving this inequality implies $n \leq 4$ which is a contradiction. □

Chapter 13

Generating Functions

We already saw in Chapter 10 that when solving a counting problem with a free parameter n, it is often helpful to think of the sequence (a_n) of all the answers. A generating function is another way to look at all these numbers simultaneously, by forming the function $f(X) = \sum_n a_n X^n$. Generating functions can also be used to record sets. For example, if A is a subset of the non-negative integers, then we can form the generating function $f(X) = \sum_{a \in A} X^a$.

At first glance this may not seem very helpful, we have only rewritten the sequence in a different way. However, as you will soon learn in doing mathematics, rewriting things in a different way is often very helpful. In this case, we actually know quite a bit about functions (and if you have not already seen them, calculus and differential equations will teach you much more about them). One of the formulas you probably already know, which we will use extensively below is the infinite geometric series

$$\frac{1}{1-qX} = \sum_{n=0}^{\infty} q^n X^n.$$

We can also multiply functions. If $f(X) = \sum_n a_n X^n$ and $g(X) = \sum_n b_n X^n$, then

$$f(X) \cdot g(X) = \sum_n \left(\sum_k a_k b_{n-k} \right) X^n.$$

This formula for the product of two generating functions turns out to be incredibly powerful. Suppose a_n is the number of ways to choose n (unordered) objects of type A and b_n is the number of ways to choose n (unordered) objects of type B. Then the coefficient $c_n = \sum_k a_k b_{n-k}$ of the product is the number of ways to choose n unordered objects of type A or B. Using this an expert in generating functions can sometimes write down the generating function for a complicated sounding problem almost instantly. This is particularly true for problems involving partitions of the number n.

Another thing one can do with functions is to take the derivative. If $f(X) = \sum_n a_n X^n$ is a function, then its derivative is $f'(X) = \sum_n n a_n X^{n-1}$. The full power of derivatives is the subject of calculus, we will be content with just one basic identity, the product rule. If f and g are functions, then

$$(f \cdot g)' = f' \cdot g + f \cdot g'.$$

Combining this with induction, gives a similar product rule for products of any number of factors.

There is one subtle detail we have not discussed. How should we interpret the X in our generating functions? In practice, we can always get by treating X as a sufficiently small number. For example, in the geometric series above, everything works out if $|X| < 1/q$. Then the function on the left exists, the series on the right converges, and the two are numerically equal. However, one should be aware that there are examples where this is not good enough. The generating function for the sequence of factorials, $f(X) = \sum_{n=0}^{\infty} n! X^n$ does not make sense for any nonzero X. In some sense one has to treat X as infinitesimally small.

However you interpret X, the most important fact is that if two generating functions agree $\sum_n a_n X^n = \sum_n b_n X^n$, then all their coefficients agree, $a_n = b_n$ for all n. We will often use this fact to extract a formula for a_n from its generating function. Occasionally, we will use it to show that two counts give the same answer, even though we will not necessarily give a formula for this common answer.

Finally, let us note that we are just presenting the tip of an iceberg here. The generating functions we are using are sometimes called ordinary generating functions. There are many other forms of generating functions, each with their own area of application.

Example 116. Find the general term of the sequence

$$a_{n+3} = -4a_{n+2} - a_{n+1} + 6a_n, \; n \geq 0$$

with the initial conditions $a_0 = 1$, $a_1 = a_2 = 2$.

Solution. We attach the generating function $f(X) = \sum_{n \geq 0} a_n X^n$. We have that

$$f(X) = a_0 + a_1 X + a_2 X^2 + \sum_{n \geq 0} a_{n+3} X^{n+3}$$
$$= a_0 + a_1 X + a_2 X^2 + X^3 \sum_{n \geq 0} (-4a_{n+2} - a_{n+1} + 6a_n) X^n.$$

Thus is we regroup terms carefully we note that we get the identity

$$f(X) = a_0 + a_1 X + a_2 X^2 - 4X f(X) - X^2 f(X) + 6X^3 f(X)$$
$$+ 4X(a_0 + a_1 X) + a_0 X^2.$$

All in all it follows

$$f(X) = \frac{1 + 6X + 11X^2}{1 + 4X + X^2 - 6X^3} = \frac{1 + 6X + 11X^2}{(1-X)(1+2X)(1+3X)}.$$

To finish up we use a technique which is called partial fraction decomposition which tells us that there should be some numbers A, B, C such that

$$\frac{1 + 6X + 11X^2}{(1-X)(1+2X)(1+3X)} = \frac{A}{1-X} + \frac{B}{1+2X} + \frac{C}{1+3X}.$$

It is easy to see what A, B, C should be, namely if for example we want to find A we multiply both sides by $1 - X$ and then plug in $X = 1$. Doing the computations we have $A = \frac{3}{2}$, $B = -1$, $C = \frac{1}{2}$.

Using the result (1) we get that

$$f(X) = \sum_{n \geq 0} \left(\frac{3}{2} - (-2)^n + \frac{(-3)^n}{2} \right) X^n.$$

Finally we apply the principle that if we have two representations of the same function, then the coefficients should agree.

Thus the general term is $a_n = \frac{3}{2} - (-2)^n + \frac{(-3)^n}{2}$. □

Counting problems can be also solved elegantly using generating functions.

Example 117. In how many ways can we fill a basket with n fruits if we have the following constraints:

- The number of oranges is even.

- We can have at most three bananas.

- The number of pineapples is divisible by 4.
- There can be at most one watermelon.
- All fruits are bananas, oranges, pineapples, and watermelons.

Solution. Again let f_n be the number of ways to fill the basket. Looking at the generating function $F(X) = \sum_{n \geq 0} f_n X^n$, since the basket is made from a combination of the fruits we can say that it is the product of the generating functions for choosing oranges, bananas, pineapples and watermelon.

The hypothesis that for oranges we have the number of choices to be even so the function is

$$O(X) = 1 + X^2 + \ldots + X^{2n} + \ldots = \frac{1}{1-X^2}.$$

For bananas, it is easy we have

$$B(X) = 1 + X + X^2 + X^3 = \frac{1-X^4}{1-X}.$$

For pineapples, we have that since the number is divisible by 4, the generating function is

$$P(X) = 1 + X^4 + \ldots + X^{4n} + \ldots = \frac{1}{1-X^4}.$$

Finally for watermelons, we have $W(X) = 1 + X$.
Thus

$$F(X) = \frac{1}{1-X^2} \cdot \frac{1-X^4}{1-X} \cdot \frac{1}{1-X^4} \cdot (1+X) = \frac{1}{(1-X)^2}.$$

Now we just need to know what is the power series expansion for $\frac{1}{(1-X)^2}$.
For this we introduce the generalized binomial coefficient, namely

$$\binom{\alpha}{n} = \frac{\alpha(\alpha-1)\ldots(\alpha-n+1)}{n!}$$

for any α. Then we have the generalized binomial identity

$$(1+X)^\alpha = \sum_{n \geq 0} \binom{\alpha}{n} X^n. \tag{2}$$

Now substitute in this relation $\alpha = -2$ and instead of X put $-X$ and we get

$$\frac{1}{(1-X)^2} = \sum_{n \geq 0} (n+1) X^n.$$

Thus $f_n = n+1$ so there are $n+1$ ways to fill the basket. □

Generating functions are very useful in showing combinatorial identities and computing sums. We begin first with a computational theorem that is going to be used in problems.

Theorem 7.
$$\sum_{n\geq 0}\binom{n}{k}X^n = \frac{X^k}{(1-X)^{k+1}} \qquad (3)$$

Proof. We have that
$$\sum_{n\geq 0}\binom{n}{k}X^n = \sum_{n\geq 0}\binom{n+k}{k}X^{n+k} = X^k\sum_{n\geq 0}\binom{n+k}{k}X^n.$$

Next we see that
$$\binom{n+k}{k} = \frac{(k+1)(k+2)\ldots(k+n)}{n!}$$
$$= (-1)^n\frac{(-(k+1))(-(k+1)-1)\ldots(-(k+1)-n+1)}{n!} = (-1)^n\binom{-(k+1)}{n}.$$

Thus
$$\sum_{n\geq 0}\binom{n}{k}X^n = X^k\sum_{n\geq 0}\binom{-(k+1)}{n}(-X)^n = \frac{X^k}{(1-X)^{k+1}}. \qquad \square$$

Example 118. Let $n \in \mathbb{N}$ and consider all polynomials with coefficients $\{0, 1, 2, 3\}$. How many of these satisfy $P(2) = n$?
(All Soviet Mathematical Olympiad, 1986)

Solution. We let a_n be the number of such polynomials and we will try to write down a formula for $f(X) = \sum_{n\geq 0} a_n X^n$.

First, suppose we restrict to polynomials of degree at most d. Then the corresponding generating function would be
$$f_d(X) = \sum_P X^{P(2)},$$
where the sum runs over all polynomials $P(x) = c_0 + c_1 x + \cdots + c_d x^d$ with each $c_k \in \{0, 1, 2, 3\}$. We recognize this as a $(d+1)$-fold sum over the possibilities of a single coefficient, and then we recognize this sum as a product of $d+1$ factors. This gives
$$f_d(X) = \sum_{c_0=0}^{3}\sum_{c_1=0}^{3}\cdots\sum_{c_d=0}^{3} X^{c_0+2c_1+\cdots+2^d c_d} = \prod_{k=0}^{d}\left(1 + X^{2^k} + X^{2\cdot 2^k} + X^{3\cdot 2^k}\right).$$

Now we use the identity $\dfrac{1-a^4}{1-a} = 1 + a + a^2 + a^3$ to obtain that

$$f_d(X) = \prod_{k=0}^{d} \frac{1-X^{2k+2}}{1-X^{2k}} = \frac{(1-X^{2^{d+1}})(1-X^{2^{d+2}})}{(1-X)(1-X^2)}.$$

Now, every coefficient of $f_d(X)$ for exponents smaller than $X^{2^{d+1}}$ agrees with the corresponding coefficient of $f(X)$. Hence as we let d grow, the left hand side becomes $f(X)$. For the right hand side, if X is small enough ($|X| < 1$ suffices), then the numerator tends to 1. Thus

$$f(X) = \frac{1}{(1-X)(1-X^2)} = \frac{1}{(1-X)^2(1+X)}.$$

We use again the partial fraction decomposition, so we must find A, B, C such that

$$\frac{1}{(1-X)^2(1+X)} = \frac{A}{1-X} + \frac{B}{(1-X)^2} + \frac{C}{1+X}.$$

After solving it we find $A = \dfrac{1}{4}$, $B = \dfrac{1}{2}$, $C = \dfrac{1}{4}$.

From the previous problem we know the expansion for $\dfrac{1}{(1-X)^2}$.

Putting it all together we have for our problem

$$f(X) = \sum_{n \geq 0} \left(\frac{1}{4} + \frac{(-1)^n}{4} + \frac{n+1}{2} \right) X^n.$$

Thus we get

$$a_n = \left[\frac{n}{2} \right] + 1. \qquad \square$$

Example 119. We throw a regular die (which gives a number in $\{1, 2, \ldots, 6\}$) n times. What fraction of the possible rolls has the sums of the outcomes divisible by 5?

(IMC 1999)

Solution. For an arbitrary positive integer k let p_k the probability that the sum of the outcomes is k. We consider the generating function $f(X) = \sum_{k \geq 1} p_k X^k$.

We claim that we have the identity

$$\sum_{k \geq 1} p_k X^k = \left(\frac{X + X^2 + X^3 + X^4 + X^5 + X^6}{6} \right)^n.$$

Generating Functions

This can be either seen using induction or we can reason that if $k = c_1 + \ldots + c_n$ with these numbers in the set $\{1, 2, \ldots, 6\}$ then the contribution to X^k is $\dfrac{1}{6^n}$ and we notice that $\dfrac{X^k}{6^n} = \prod_{i=1}^{n} \dfrac{X^{c_i}}{6}$.

What we are interested in finding is $\sum_{k} p_{5k}$. Here comes into play the following useful trick, namely take ω a fifth root of unity.

The usefulness of this is the fact that

$$f(1) + f(\omega) + f(\omega^2) + f(\omega^3) + f(\omega^4) = 5\sum_{k} p_{5k}.$$

Now for $j \in \{1, 2, 3, 4\}$ we have $\omega^j + \omega^{2j} + \omega^{3j} + \omega^{4j} + \omega^{5j} = 0$.
Thus we get

$$5\sum_{k} p_{5k} = 1 + \dfrac{1}{6^n}(\omega^n + \omega^{2n} + \omega^{3n} + \omega^{4n}).$$

To sum up we get that the probability is $\dfrac{1}{5} - \dfrac{1}{6^n}$ for n not divisible by 5 and otherwise it is $\dfrac{1}{5} + \dfrac{4}{6^n}$. For an arbitrary positive integer k, let p_k the fraction of rolls for which the sum of the outcomes is k. We consider the generating function $f(X) = \sum_{k \geq 1} p_k X^k$. We claim that we have the identity

$$\sum_{k \geq 1} p_k X^k = \left(\dfrac{X + X^2 + X^3 + X^4 + X^5 + X^6}{6}\right)^n$$

This can be either seen using induction or we can reason as follows. Record a sequence of rolls as a string $(c_1, c_2, \ldots, c_n)$ where $1 \leq c_j \leq 6$. Then

$$k = c_1 + c_2 + \cdots + c_n) \quad \text{and} \quad \dfrac{X^k}{6^n} = \prod_{i=1}^{n} \dfrac{X^{c_i}}{6},$$

so we recognize the sum over all possible sequences of rolls as a product of n identical terms

$$f(X) = \sum_{c_1=1}^{6} \sum_{c_2=1}^{6} \cdots \sum_{c_n=1}^{6} \dfrac{1}{6^n} X^{c_1+c_2+\cdots+c_n} = \left(\sum_{c=1}^{6} \dfrac{1}{6} X^c\right)^n.$$

What we are interested in finding is $\sum_{k} p_{5k}$. Here the following useful trick comes into play, namely plug in $X = \omega$, a fifth root of unity. The reason this is useful is the fact that

$$f(1)+f(\omega)+f(\omega^2)+f(\omega^3)+f(\omega^4) = \sum_n (1+\omega^n+\omega^{2n}+\omega^{3n}+\omega^{4n})p_n = 5\sum_k p_{5k},$$

where we have noticed that $1+\omega^n+\omega^{2n}+\omega^{3n}+\omega^{4n}$ is 5 if $n = 5k$ is a multiple of 5, and is zero otherwise. This last identity also lets us compute $f(\omega^j)$ easily. For $j \in \{1, 2, 3, 4\}$ we have $\omega^j + \omega^{2j} + \omega^{3j} + \omega^{4j} + \omega^{5j} = 0$. Hence

$$f(\omega^j) = \left(\frac{\omega^{6j}}{6}\right)^n = \frac{\omega^{jn}}{6^n}.$$

Thus we get

$$5\sum_k p_{5k} = 1 + \frac{1}{6^n}(\omega^n + \omega^{2n} + \omega^{3n} + \omega^{4n}).$$

To sum up, we get that the fraction is $\frac{1}{5} - \frac{1}{6^n}$ for n not divisible by 5 and otherwise it is $\frac{1}{5} + \frac{4}{6^n}$. □

Next let us show how generating functions can be used in dealing with partitions of the natural numbers.

Example 120. For a set S of nonnegative integers, let $r_S(n)$ denote the number of ordered pairs (s_1, s_2), $s_1, s_2 \in S$ and $s_1 \neq s_2$ such that $n = s_1 + s_2$. Is it possible to partition the natural numbers into two sets A and B such that $r_A(n) = r_B(n)$ for all n?

(Putnam 2003)

Solution. The answer is "yes". One could cleverly guess the correct partition and prove that it works. However, we will show how to use generating functions to find the partition. Assume without loss that $0 \in A$ and consider the generating function of set A, $f(X) = \sum_{a \in A} X^a$ and the generating function of set B, $g(X) = \sum_{b \in B} X^b$.

Now the problem states that we should be looking at sums of the type $a_i + a_j$ with $a_i \neq a_j$. Now this type of sum occurs naturally if we take $f^2(X)$. More explicitly, we have that

$$f^2(X) = \sum_{a \in A} X^{2a} + 2\sum_{n \in \mathbb{N}} r_A(n)X^n = f(X^2) + 2\sum_{n \in \mathbb{N}} r_A(n)X^n.$$

Similarly,

$$g^2(X) = \sum_{b \in B} X^{2b} + 2\sum_{n \in \mathbb{N}} r_B(n)X^n = g(X^2) + 2\sum_{n \in \mathbb{N}} r_B(n)X^n.$$

Thus the problem states that we should have

$$f^2(X) - f(X^2) = g^2(X) - g(X^2).$$

We can rewrite this as

$$f^2(X) - g^2(X) = f(X^2) - g(X^2)$$

or

$$(f(X) - g(X))(f(X) + g(X)) = f(X^2) - g(X^2).$$

Now A and B form a partition of the natural numbers so

$$f(X) + g(X) = \sum_{n \in \mathbb{N}} X^n = \frac{1}{1-X}.$$

Thus, if we denote with $h(X) = f(X) - g(X)$, we have that

$$h(X) = (1-X)h(X^2).$$

Iterating this procedure, we find that

$$h(X) = \prod_{i=1}^{n} \left(1 - X^{2^i}\right) h(X^{2^n}).$$

If $|X|$ is small (and as usual $|X| < 1$ suffices), then letting n get large $h(X^{2^n})$ will tend to $h(0) = f(0) - g(0) = 1 - 0 = 1$ (since $0 \in A$). Hence

$$h(X) = \prod_{i=1}^{\infty} \left(1 - X^{2^i}\right).$$

Thus there is only one possibility for what h could be. If we expand this out, we will indeed see that every coefficient is either 1 or -1. The coefficient of X^n is 1 if the binary expansion of n has an even number of ones (and therefore these numbers form A), and it is -1 if the binary expansion has an odd number of ones (and these numbers form B).

Alternatively, if we write $h = \sum_{n \in \mathbb{N}} c_i X^i$, then the relation

$$h(X) = (1-X)h(X^2)$$

can be restated as $c_{2i} = c_i$ and $c_{2i+1} = -c_i$. So if we start with $c_0 = 1$, then it is easy to see that defining the coefficients by this recursion we get a sequence with values -1 and 1. □

Example 121. Let n be a positive integer. Let $d(n)$ denote the number of partitions of n with distinct parts and let $o(n)$ equal the number of partitions of n with odd parts. Prove that $d(n) = o(n)$.

Solution. Again the motto is that we want to prove such an equality then it suffices to show that the two sequences have the same generating function.

Let $D(X) = \sum_{n \geq 1} d(n) X^n$ and $O(n) = \sum_{n \geq 1} o(n) X^n$.

Looking at how we define these sequences we see that

$$D(X) = \prod_{j=1}^{\infty}(1 + X^j)$$

and

$$O(X) = \prod_{k=1}^{\infty}(1 + X^{2k+1} + X^{2(2k+1)} + \ldots + X^{n(2k+1)} + \ldots) = \prod_{k=1}^{\infty} \frac{1}{1 - X^{2k+1}}.$$

Next we can rearrange the terms in $D(X)$ as

$$D(X) = \prod_{k=1}^{\infty} \prod_{i=1}^{\infty} (1 + X^{2^i(2k+1)}).$$

To finish let us note that

$$\prod_{i=1}^{\infty}(1 + X^{2^i(2k+1)}) = \frac{1}{1 - X^{2k+1}}$$

since for any l we have that

$$(1 - X^{2k+1}) \prod_{i=1}^{l}(1 + X^{2^i(2k+1)}) = 1 - X^{2^{l+1}(2k+1)}$$

and letting l go to infinity, and using the fact that $|X| < 1$ we obtain that the term $1 - X^{2^{l+1}(2k+1)}$ will converge to 1 and we are done. $\square$

The general principle if we want to establish combinatorial identities is to prove that the two quantities stated in a given problem have the same generating function.

Example 122. Prove that $\sum_{k=0}^{m} \binom{m}{k} \binom{n+k}{m} = \sum_{k=0}^{m} \binom{m}{k} \binom{n}{k} 2^k$, using the generalized binomial identity (2).

Solution. We must prove

$$\sum_{n=0}^{\infty}\sum_{k=0}^{m}\binom{m}{k}\binom{n+k}{m}X^n = \sum_{n=0}^{\infty}\sum_{k=0}^{m}\binom{m}{k}\binom{n}{k}2^k X^n.$$

For the left hand side, we have

$$\sum_{n=0}^{\infty}\sum_{k=0}^{m}\binom{m}{k}\binom{n+k}{m}X^n = \sum_{k=0}^{m}\binom{m}{k}\frac{1}{X^k}\sum_{n=0}^{\infty}\binom{n+k}{m}X^{n+k}$$

$$\stackrel{(3)}{=} \sum_{k=0}^{m}\binom{m}{k}\frac{1}{X^k}\frac{X^m}{(1-X)^{m+1}}$$

$$= \frac{X^m}{(1-X)^{m+1}}\sum_{k=0}^{m}\binom{m}{k}\frac{1}{X^k}$$

$$= \frac{X^m}{(1-X)^{m+1}}\left(1+\frac{1}{X}\right)^m$$

$$= \frac{(X+1)^m}{(1-X)^{m+1}}.$$

For the right hand side, we have

$$\sum_{n=0}^{\infty}\sum_{k=0}^{m}\binom{m}{k}\binom{n}{k}2^k X^n = \sum_{k=0}^{m}\binom{m}{k}2^k\sum_{n=0}^{\infty}\binom{n}{k}X^n$$

$$= \sum_{k=0}^{m}\binom{m}{k}2^k\frac{X^k}{(1-X)^{k+1}}$$

$$= \frac{1}{1-X}\sum_{k=0}^{m}\binom{m}{k}\left(\frac{2X}{1-X}\right)^k$$

$$= \frac{1}{1-X}\left(1+\frac{2X}{1-X}\right)^m$$

$$= \frac{(X+1)^m}{(1-X)^{m+1}}.$$

This finishes the problem since both sides have the same generating function. $\square$

Example 123. Find $\sum_{\substack{a_1+\ldots+a_k=n \\ a_1,a_2,\cdots,a_k>0}} a_1 a_2 \ldots a_k$, where n is a natural number and the sum is taken over all partitions of it.

Solution. Let s_n be the sum in question. We want to determine the generating function of his sequence, namely $f(X) = \sum_n s_n X^n$. Observe that we can rewrite f as

$$f(X) = \sum_{k \geq 0} \sum_{n \geq 0} \sum_{\substack{a_1 + \ldots + a_k = n \\ a_1, a_2, \cdots, a_k > 0}} a_1 \ldots a_k X^{a_1 + \ldots + a_k}$$

$$= \sum_{k \geq 0} \sum_{a_1 > 0} \sum_{a_2 > 0} \cdots \sum_{a_k > 0} a_1 \ldots a_k X^{a_1 + \ldots + a_k}$$

$$= \sum_{k \geq 0} \left(\sum_a a X^a \right)^k.$$

Now if we use Theorem 7 for $k = 1$, we get that

$$\sum_{i \geq 0} i X^i = \frac{X}{(1-X)^2}.$$

So our generating function is

$$f(X) = \sum_{k \geq 0} \frac{X^k}{(1-X)^{2k}} = \sum_{k \geq 0} (X(1-X)^{-2})^k$$

$$= \frac{1}{1 - X(1-X)^{-2}} = \frac{(1-X)^2}{(1-X)^2 - X}$$

$$= 1 + \frac{X}{X^2 - 3X + 1}.$$

We can use partial fractions to extract an explicit formula for s_n. Since we want to use the infinite geometric series, we factor the denominator as $(1 - r_1 X)(1 - r_2 X)$. This means we need $1/r_1$ and $1/r_2$ to be the roots of $X^2 - 3X + 1 = 0$. Hence r_1 and r_2 are the roots of the reversed polynomial (which in this case is the same by symmetry) $t^2 - 3t + 1 = 0$. Thus we find

$$r_1 = \frac{3 + \sqrt{5}}{2} \quad \text{and} \quad r_2 = \frac{3 - \sqrt{5}}{2}.$$

Either by the usual trick of multiplying both sides by $1 - r_j X$ and plugging in $X = 1/r_j$ or by the sneaky trick of rewriting the numerator as

$$X = \frac{(1 - r_2 X) - (1 - r_1)X}{r_1 - r_2},$$

we get

$$\frac{X}{X^2 - 3X + 1} = \frac{X}{(1 - r_1 X)(1 - r_2 X)}$$

$$= \frac{1}{r_1 - r_2} \cdot \frac{1}{1 - r_1 X} - \frac{1}{r_1 - r_2} \cdot \frac{1}{1 - r_2 X}.$$

Thus $f(X) = 1 + \sum_{k=0}^{\infty} \frac{r_1^k - r_2^k}{r_1 - r_2} \cdot X^k$, and we obtain that $s_n = \frac{r_1^n - r_2^n}{r_1 - r_2}$.

You can get a better formula than this if you note that

$$r_1 = \left(\frac{1+\sqrt{5}}{2}\right)^2 \quad \text{and} \quad r_2 = \left(\frac{1-\sqrt{5}}{2}\right)^2$$

are the squares of the roots of $t^2 - t + 1$, the characteristic polynomial for the Fibonacci recursion. Then we find

$$s_n = \frac{1}{\sqrt{5}}\left(\left(\frac{1+\sqrt{5}}{2}\right)^{2n} - \left(\frac{1-\sqrt{5}}{2}\right)^{2n}\right) = F_{2n}.$$

The reader is encouraged to reprove this last formula using the generating function $\frac{X}{1-X-X^2}$ for the Fibonacci numbers. □

Example 124. Let $n \geq 3$ a positive integer. We say that a subset $A \subset \{1, 2, \ldots, n\}$ is even if the sum of its elements is even. Otherwise we call that subset odd. By convention, we define the sum of the empty set to be 0, so it is even.

a) How many even and odd subsets does $\{1, 2, \ldots, n\}$ have?

b) Find the sum of the elements of the even, respectively of the odd subsets of $\{1, 2, \ldots, n\}$.

(Romania, TST 1994)

Solution. Consider the generating function

$$f(X) = \prod_{k=1}^{n}(1 + X^k).$$

If we expand it out, we can write it as

$$f(X) = \sum_{A \subset \{1,2,\ldots,n\}} X^{\sigma(A)},$$

where we define $\sigma(A)$ to be the sum of elements of the subset A of $\{1, 2, \ldots, n\}$.

Next, we see how we detect even and odd sums. The idea is to take $X = -1$, then $(-1)^{\sigma(A)} = 1$ if A is an even subset and -1 if A is an odd subset. Thus $f(-1)$ gives us the difference between the number of even and odd subsets. Since $f(-1) = 0$, we have the numbers are the same, namely 2^{n-1}.

Now for part b), we need to find sums of elements and the way to do it is to notice that if you have $X^{\sigma(A)}$ there is and easy way to obtain $\sigma(A)$, namely take derivative. Using this we have that

$$f'(X) = \sum_{A \subset \{1,2,\ldots,n\}} \sigma(A) X^{\sigma(A)-1}.$$

On the other hand, since f is a product of n factors, we can obtain its derivative from the product formula. From the product formula for a product of two factors and induction, we see that the derivative of a product of n terms will be a sum of n terms, where in each term we take the derivative of one of the factors. In this case we can write this as

$$f'(X) = \prod_{k=1}^{n}(1+X^k) \cdot \sum_{k=1}^{n} \frac{kX^{k-1}}{1+X^k}.$$

Denoting by E the sum of elements of even subsets and by O the sum of elements of odd subsets, we have that $f'(1) = O + E$ and $f'(-1) = O - E$. It is easy to compute $f'(1) = 2^{n-1}(1+2+\ldots+n) = n(n+1)2^{n-2}$. To compute $f'(-1)$, note that, for $n \geq 3$, there are at least two factors of f that vanish at $X = -1$, hence after using the product rule every summand still has at least one term that vanishes at $X = -1$ and hence $f'(-1) = 0$. Thus we obtain $E = O = n(n+1)2^{n-3}$. □

Example 125. Out of the 90-element subsets of $\{1, 2, \ldots, 2014\}$ are there more with an even sum of elements or more with an odd sum of elements? Moreover, find a precise count.

Solution. The prototype of the generating function we should consider is $\prod_{j=1}^{2014}(1 + y^j X)$. We will use y to filter the even and odd sums, by setting it to 1 and -1. The key is to notice that the coefficient of X^{90} in this product is equal to

$$\sum_{1 \leq j_1 < j_2 < \ldots < j_{90} \leq 2014} y^{j_1 + j_2 + \ldots j_{90}}.$$

Let A be the number of subsets with even sum and B the number of subsets with odd sum. Then if we plug in $y = 1$, the coefficient of X^{90} will be $A + B$ and if we plug in $y = -1$, the coefficient of X^{90} will be $A - B$. We just need to compute these coefficients.

If we plug in $y = 1$, the product becomes $(1 + X)^{2014}$, and we read off the coefficient of X^{90} as $A + B = \binom{2014}{90}$.

Generating Functions

If we plug in $y = -1$, the product becomes

$$\prod_{j=1}^{2014}(1 + (-1)^j X) = (1+X)^{1007}(1-X)^{1007} = (1-X^2)^{1007},$$

since there 1007 odd numbers and 1007 even numbers between 1 and 2014. The X^{90} term in this polynomial is

$$A - B = \binom{1007}{45}(-X^2)^{45} = -\binom{1007}{45}X^{90}.$$

Thus $B - A = \binom{1007}{45} > 0$ so there are more subsets with an odd sum of elements. Adding and subtracting these two formulas gives

$$B = \frac{1}{2}\left(\binom{2014}{90} + \binom{1007}{45}\right) \quad \text{and} \quad A = \frac{1}{2}\left(\binom{2014}{90} - \binom{1007}{45}\right). \quad \square$$

Finally it is time for a difficult problem. This was the sixth problem on the IMO contest that year, and this solution given by Nikolai Nikolov, received a special prize. It uses the usual trick of plugging in a root of unity, but with a clever approach that simplifies the argument.

Example 126. Fix a prime $p > 2$. Find the number of p-element subsets of $\{1, 2, \ldots, 2p\}$ with sum of the elements divisible by p.

(IMO 1996)

Solution. Let us look again at the generating function $\prod_{j=1}^{2p}(X - y^j)$. Now if we look at the coefficient of X^p this is equal to

$$-\sum_{1 \leq j_1 < j_2 < \ldots < j_p \leq 2p} y^{j_1 + j_2 + \ldots j_p}.$$

We see that we need something that detects sums modulo p in the exponent. The gadget that does this is the p-th root of unity call it ε.

If we denote with a_i the number of p-element subsets with sum congruent to i modulo p, for $0 \leq i \leq p-1$, then we get that coefficient of X^p in $\prod_{j=1}^{2p}(X - \varepsilon^j)$ is equal to

$$-(a_0 + a_1 \varepsilon + \ldots + a_i \varepsilon^i + \ldots + a_{p-1}\varepsilon^{p-1}).$$

On the other hand we have that

$$\prod_{j=1}^{2p}(X - \varepsilon^j) = \left((X-1)(X-\varepsilon)\ldots(X-\varepsilon^{p-1})\right)^2 = (X^p - 1)^2 = X^{2p} - 2X^p + 1.$$

Thus
$$a_0 + a_1\varepsilon + \ldots + a_i\varepsilon^i + \ldots + a_{p-1}\varepsilon^{p-1} = 2.$$

This means that ε is a root of the polynomial
$$a_0 - 2 + a_1 X + \ldots + a_{p-1}X^{p-1}.$$

The minimal polynomial of ε is $X^{p-1} + X^{p-2} + \ldots + X + 1$ and this implies
$$X^{p-1} + X^{p-2} + \ldots + X + 1 | a_0 - 2 + a_1 X + \ldots + a_{p-1}X^{p-1}.$$

Since these two polynomials have the same degree it follows that there is a λ such that
$$\lambda(X^{p-1} + X^{p-2} + \ldots + X + 1) = a_0 - 2 + a_1 X + \ldots + a_{p-1}X^{p-1}.$$

We obtain that $a_0 - 2 = \lambda$ and $a_i = \lambda$, for $i \geq 1$. Finally note that for every p-element subset of $\{1, 2, \ldots, 2p\}$ we have some residue modulo p for the sum of elements, so
$$a_0 + a_1 + \ldots + a_p = \binom{2p}{p}.$$

Putting it together, we get that
$$a_0 = \frac{\binom{2p}{p} - 2}{p} + 2. \qquad \square$$

Chapter 14

Probabilities and Probabilistic Method

Probabilities provide another powerful way of reworking combinatorial problems. They provide a different language for doing counts. However some of the most amazing applications of probability to combinatorics arise because it provides a powerful extension of the Pigeonhole Principle.

We will only be concerned with probabilities on a finite set S, called the sample space. In this case we can define a probability as follows.

Definition. A probability is a function $P : S \to [0,1]$ such that $\sum_{s \in S} P(s) = 1$.

In probability, a subset A of S is referred to as an event. The probability of an event is given by $P(A) = \sum_{s \in A} P(s)$. Note that by definition $P(S) = 1$. In most of our examples all $s \in S$ have the same probability. In this case we simply have $P(A) = \dfrac{|A|}{|S|}$. Thus computing the probability of A is essentially the same as counting the elements of A.

For example, if we are rolling a standard die, then the natural sample space is $S = \{1, 2, 3, 4, 5, 6\}$, where we interpret $s \in S$ as meaning we roll the die and see s dots on the top face. If the die is fair, then each element of S will have probability $1/6$. If we let $A = \{1, 3, 5\}$, then we can think of A as the event that we see an odd number of dots on the top face, and $P(A) = 1/2$.

All the ideas from counting extend to probabilities. If A is an event, then the complementary event, denoted A', is the complement of A in S. Then we have $P(A') = 1 - P(A)$. If A and B are two events, then the simplest case of Inclusion-Exclusion becomes the formula $P(A \cap B) + P(A \cup B) = P(A) + P(B)$.

One of the most important notions in probability is independence. We say two events A and B are independent if $P(A \cap B) = P(A)P(B)$.

To understand this definition, suppose you are told that A happened. Then the probability that B also happens in $P(A \cap B)$. So the fraction of the time that B happens, given that A happened, is $\dfrac{P(A \cap B)}{P(A)} = P(B)$. Hence telling you that A happens, gives you no information about whether B happened.

One of the most important applications of independence is that it lets us quickly build large sample spaces and probabilities on them. We can say something like: Roll a fair die 7 times, different rolls are independent. This means that the sample space is $\{1, 2, 3, 4, 5, 6\}^7$, with each factor corresponding to one of the rolls. If A is any set defined by fixing some coordinates and B is any set defined by fixing a disjoint set of coordinates, then A and B are independent. In this simple case, this just means each element has probability $1/6^7$.

A function defined on S is called a random variable. For us, all random variables will have real values. In probability, random variables are usually denoted capital letters and often X, Y, or Z. Particular values that these random variables might take on are often denoted by the corresponding lower case letters. This notation arises because we might imagine doing some complicated experiment (roll a fair die 7 times) and reading off some number (the sum of the 7 rolls). We use X to talk about this quantity in the abstract, but x to talk about the value it took on in an actual instance of the experiment.

If X is a random variable on a sample space S, then we can also talk about probabilities for X, $P(X = x) = \sum_{s \in S : X(s) = x} P(s)$, where the sum is taken over all elements $s \in S$ for which the random variable X takes on the value x. We can also talk about independence, two random variables X and Y are independent, if for all x and y, we have

$$P(X = x \text{ and } Y = y) = P(X = x)P(Y = y).$$

The most important definition for random variables is their expected value.

Definition. The expected value of random variable X is given by

$$E[X] = \sum_{s \in S} X(s) P(s) = \sum_{x} x P(X = x),$$

where in the second sum x ranges over all possible values of X.

In a technical sense that we cannot really get into here, if we repeated our experiment many times and recorded the values of the random variable for each of them, then the average of these observed values would be close to the expected value. The most important property of expected values is their linearity. If X and Y are random variables, then $E[X + Y] = E[X] + E[Y]$. The reader will recognize this as counting in two ways, just translated into probability language.

Probabilities and Probabilistic Method 139

Let's start working through some examples and as we go along we will point out useful strategies.

Example 127. There are 30 students in class. What is the probability that two students share the same birthday? Assume that each of the 365 days are equally likely and different birthdays are independent.

Solution. It is easier to count the complementary event; namely all students having different birthdays. We consider a random order of students. Take the first student, and mark his birthday. Now move to the second; the probability that it has a different birthday from the first is $1 - \frac{1}{365}$ and also mark it. Look at the third and now you have two marks in the calendar and thus the probability of a new mark is $1 - \frac{2}{365}$. Thus in general we get for the $k+1$-th student a probability of $1 - \frac{k}{365}$ that his birthday does not match one already marked.

Since different birthdays are independent, the probability that all students have different birthdays is the product of these individual probabilities. Thus we get this probability is equal to $\prod_{k=1}^{29} \left(1 - \frac{k}{365}\right)$. If we compute this we get this is less than 0.3 and thus the probability of two students sharing the same birthday is at least 0.7. □

Example 128. Suppose you have a fair coin. What is the probability that you need to flip it n times until you obtain a head? Now imagine that you are paid n dollars if you get a head after exactly n flips. What is the expected payment you receive?

Solution. Note that since the coin is fair the probability of getting heads or tail is $\frac{1}{2}$ and flips are independent. To need n flips to get a head, the first $n-1$ throws you have to get tails and then hit a head. Using independence we obtain the probability to be $\frac{1}{2^n}$.

The second part can be translated into what is the "expected number of flips". Having the probability from above, we have to compute

$$E[F] = \sum_{n=1}^{\infty} \frac{n}{2^n}.$$

We know from the generating function chapter, that

$$\sum_{n=1}^{\infty} n X^n = \frac{X}{(1-X)^2}.$$

Plugging in $X = \dfrac{1}{2}$, we obtain that $E[F] = 2$. □

Example 129. Shanille O'Keal shoots free throws on a basketball courts. She hits the first and misses the second, and thereafter the probability that she hits the next shot is equal to the proportion of shots she has hit so far. What is the probability she hits exactly 50 of her first 100 shots?

(Putnam 2002)

Solution. We show by induction on n that after n shots, the probability of having made any number of shots from 1 to $n-1$ is equal to $\dfrac{i}{n-1}$. This is obvious for $n = 2$. Given the result for n, then for $n+1$ we see that the probability of making i shots is split up in two parts

- She makes $i-1$ shots in the first n attempts and makes the $n+1$-th shot. This gives us $\dfrac{i-1}{n} \cdot \dfrac{1}{n-1}$.

- She makes i shots in the first n attempts and misses the $n+1$-th shot. This gives us $\left(1 - \dfrac{i}{n}\right) \cdot \dfrac{1}{n-1}$.

Summing up we get $\left(1 - \dfrac{1}{n}\right) \cdot \dfrac{1}{n-1} = \dfrac{1}{n}$, and the induction is complete.

Thus the probability in our case is equal to $\dfrac{1}{99}$. □

Example 130. Show that among 2^{100} people, there do not necessarily exist 200 people who know each other, or 200 people such that no two are acquainted.

Solution. This is our first example of using probabilities to prove a Pigeonhole Principle-like result. The first trick here is to build the correct probability space.

If we had $n = 2^{100}$ people, then we could record which pairs are acquainted by a graph, with n vertices corresponding to the people and an edge between two people if they are acquainted. Hence we want to describe a probability space for building random graphs.

Here is an easy way to do this, which works for this problem. Draw the n vertices. For each edge, we include the edge with probability $\dfrac{1}{2}$ with different edges independent. (You could imagine going through the graph one edge at a time. For each edge you flip a fair coin and you include the edge if you get heads and omit it if you get tails.)

Now let X be the random variable whose value is the number of $k = 200$ element subsets for which either all k vertices are pairwise adjacent (the corresponding k people are all pairwise acquainted) or are all pairwise not adjacent (the corresponding k people are all pairwise not acquainted). For a particular k element subset A, let X_A be the random variable that is 1 if those k elements are all pairwise adjacent or all pairwise not adjacent. Then $X = \sum_A X_A$ and hence $E[X] = \sum_A E[X_A]$. This sum has $\binom{n}{k}$ terms. The probability that all k vertices of A pairwise adjacent is $\dfrac{1}{2^{\binom{k}{2}}}$, since we are specifying the outcome for $\binom{k}{2}$ flips. The probability that they are all pairwise not adjacent is the same. Hence $E[X_A] = 2^{1-\binom{k}{2}}$. Thus $E[X] = \binom{n}{k} 2^{1-\binom{k}{2}}$.

The final step is to note that if this expected value is less than 1, then there must be some arrangement for which $X < 1$. But the values of X are nonnegative integers, so this means $X = 0$. But an arrangement with $X = 0$ is exactly what we wanted to show exists. The required computation is easy, we need to show that

$$\binom{2^{100}}{200} < \frac{(2^{100})^{200}}{200!}$$

is smaller than $2^{100 \cdot 199 - 1}$. Thus we just need to show $200! > 2^{99}$, which is obvious. □

Example 131. In airplanes, the ticket indicates, among other things, the number of the chair in which the passenger should sit. Alice, the first passenger to enter the plane, with 100 chairs, lost her ticket. So she sat in a random chair. After this, each one of the other 99 passengers sat in his chair if it was empty or in a random empty chair if his own chair was not empty.

Let $P(k)$ be the probability that the k-th passenger to enter the plane sits in his own chair. Find an expression for $P(k)$, as $2 \leq k \leq 100$.

Solution. We will solve the problem for a general number n of passengers, and we will do this by using induction on n.

We claim that $P(k) = \dfrac{n+1-k}{n+2-k}$ for $2 \leq k \leq n$. For this assume without loss of generality that passenger i has the assigned seat to be labeled with i.

For $n = 2$ the statement is clear since $P(2) = \dfrac{1}{2}$; he gets in his place, only if Alice chooses her spot.

Now for the induction step, choose a group of $n + 1$ persons, with Alice entering first. Let $2 \leq k \leq n+1$. Let a be the label of the place chosen by Alice.

Obviously $P(a = i) = \dfrac{1}{n+1}$. We have the following conditional probabilities:

- $P(k|a = k) = 0$; since if Alice picked the k-th spot there is no way for the k-th person to sit in his place

- If $a = i > k$ or $a = 1$, then each of the persons $2, \ldots, k-1$ seats in his own spot and so does the k-th person

- If $a = i$ where $2 \le i \le k - 1$. Then the passengers $2, \ldots, i-1$ can still occupy their places, but now person i will have to choose a random spot in a group of $n+1-(i-1) = n-i+2$ persons. This group has at least two persons and person number k has now the label $k-(i-1) = k-i+1 > 1$. Therefore we may use the induction hypothesis to obtain that

$$P(k|a = i) = \frac{n+2-i+1-(k-i+1)}{n+2-i+2-(k-i+1)} = \frac{n+2-k}{n+3-k}$$

Now we have by summing up the information above

$$P(k) = \sum_{i \ge 1} P(a = i) \cdot P(k|a = i)$$

$$= \frac{1}{n+1}\left[(n+2-k) + \frac{n+2-k}{n+3-k}(k-2)\right] = \frac{n+2-k}{n+3-k}. \qquad \square$$

Example 132. You have n coins $C_1, C_2, \ldots, C_n$. The i-th coin has $\dfrac{1}{(2i+1)}$ chance of coming up heads. If you flip all the coins, what is the probability that an odd number of them come up heads?

(Putnam)

Solution. If we choose $\{k_1, k_2, \ldots, k_{2r+1}\}$ to be the coins that flipped turned heads, then the rest have to turn up tails. Thus we would have to compute the following daunting sum $\sum \prod_{i=1}^{2r+1} \dfrac{1}{2k_i + 1} \prod_{j \ne k_i} \dfrac{2j}{(2j+1)}$ for all possible choice of r and subsets of $\{1, 2, \ldots, n\}$. Again we used the fact that coin flips are independent.

The idea is to go back again to a generating function; namely

$$f(X) = \prod_{i=1}^{n}\left(\frac{2i}{2i+1} + \frac{X}{2i+1}\right) = \sum_{k=0}^{n} A_k X^k$$

The daunting sum is just the of the odd indexed coefficients. Let us denote this sum with A and the sum of even indexed coefficients be B. Then $f(1) = 1 = A + B$ and $f(-1) = B - A$. Thus $A = \dfrac{f(1) - f(-1)}{2}$.

It is easy to see that $f(1) = 1$ and that $f(-1) = \prod_{k=1}^{n} \dfrac{2k-1}{2k+1} = \dfrac{1}{2n+1}$.

Thus the sought probability is $\dfrac{n}{2n+1}$. □

Example 133. In the Duma, there are 1600 members who have formed 16000 committees of 80 persons each. Prove that one can find two committees having at least four members in common.

(Russia 1996)

Solution. This is a typical situation in which the of expected value is useful. We are looking for pairs of committees and members in common. If we manage to prove that these expected values would be bigger than 3 then we win, there should two committees with at least four members in common.

So let's start setting up. We choose two committees randomly and uniformly. Let X be the random variable giving the number of people this committees have in common, and X_i the random variable saying that the i-th person is a member of both committees.

Note the obvious $X = \sum_{i=1}^{1600} X_i$ and since X_i are linearly independent we have $E[X] = \sum_{i=1}^{1600} E[X_i]$.

Now the point is that $E[X_i]$ is easy to compute. Let n_i be the number of committees of for which i is a member. Then $E[X_i] = \dfrac{\binom{n_i}{2}}{\binom{16000}{2}}$.

The sum $\sum_{i=1}^{1600} n_i = 16000 \cdot 80$, each committee is counted 80 times, for each of its members. Thus the average value of the n_i is $n = 800$.

The only thing left to do is use the convexity of the binomial function; we have
$$\sum_{i=1}^{1600} E[X_i] \geq 1600 \cdot \dfrac{\binom{n}{2}}{\binom{16000}{2}} = \dfrac{80 \cdot 799}{15999} > 3.9$$

This ends the problem since we've obtained $E[X] > 3.9$. □

Example 134. Let S be a finite set of points in the plane such that no three of them are on a line. For each convex polygon P whose vertices are in S, let $a(P)$ be the number of vertices of P, and let $b(P)$ be the number of points of S which are outside P. A line segment, a point, and the empty set are considered as convex polygons of 2, 1, and 0 vertices respectively. Prove that for every real number x:

$$\sum_P x^{a(P)}(1-x)^{b(P)} = 1,$$

where the sum is taken over all convex polygons with vertices in S.

(IMO Shortlist 2006)

Solution. Note that it suffices to prove the identity for $0 < x < 1$, since we are dealing with a polynomial.

Color the points of S white and black, with the probability that a point is colored with black equal to x and different points colored independently. Then if we look at a polygon $\mathcal{P}$, the probability that all its vertices are black and that all points outside of it are white is equal to $x^{a(P)}(1-x)^{b(P)}$.

Thus the given sum is the expected number of polygons with all vertices black and only white points outside it. Since for any set of black vertices there is precisely only one such polygon, namely its convex hull, the expected value is equal to 1. The identity thus follows. $\square$

Example 135. Several chords are chosen on a unit circle with the sum of their lengths equal to 13. Prove that there a diameter which intersects at least 5 of the chords.

Solution. This is another typical example of using probabilities, this time in a geometric setting. This will require probabilities on an infinite sample space, but we will trust the reader is comfortable enough with probability that this is not a problem.

Let $\mathcal{C}$ be the set of chords and for $c \in \mathcal{C}$, let $l(c)$ be its length.

Our sample space will be the set of diameters of the circle, with the "obvious" rotationally invariant probability distribution. Let X be the random variable which gives the number of chords cut by a diameter. For $c \in \mathcal{C}$, let X_c be the random variable that is 1 is the diameter cuts c and 0 otherwise. Then

$$X = \sum_{c \in \mathcal{C}} X_c \quad \text{and} \quad E[X] = \sum_{c \in \mathcal{C}} E[X_c].$$

Now $E[X_c]$ is just the probability that the diameter cut chord c. Viewed from the center of the circle, the chord c subtends a central angle of $2 \cdot \arcsin\left(\frac{l(c)}{2}\right)$. The set of all diameters covers a range of π in central angles, so we get that the probability of crossing c is equal to $\frac{2}{\pi} \cdot \arcsin\left(\frac{l(c)}{2}\right)$. Thus using that

$\arcsin(x) \geq x$, we have

$$E[X] = \sum_{c \in \mathcal{C}} \frac{2}{\pi} \cdot \arcsin\left(\frac{l(c)}{2}\right) \geq \sum_{c \in \mathcal{C}} \frac{2}{\pi} \cdot \frac{l(c)}{2} = \frac{1}{\pi} \sum_{c \in \mathcal{C}} l(c) > \frac{13}{\pi} \approx 4.2.$$

Thus there must be some point where $X > 4.2$, since X is integer-valued there will be will be a point where X is at least 5. $\square$

Chapter 15

Introductory Problems

1. Let k be a positive integer. In how many ways can one select a subset of three distinct numbers from the set $\{1, 2, \ldots, 3k\}$ such that their sum is divisible by 3?

2. How many n-digit numbers have their digits in non-decreasing order? (For example, the number 122379999 has its digits in non-decreasing order, but the number 12330 does not.)

3. How many North-East lattice paths are there from $(0,0)$ to $(8,8)$ that never touch the point $(4,6)$ or the point $(2,3)$?

4. A committee of 5 is to be chosen from a group of 10 people. How many ways can it be chosen if Dave and Richard must serve together or not at all, and Tina and Val refuse to serve with each other?

5. Call a number *prime-looking* if it is composite but not divisible by 2, 3, or 5. The three smallest prime-looking numbers are 49, 77, and 91. There are 168 prime numbers less than 1000. How many prime-looking numbers are there less than 1000?

(2005 AMC 12A)

6. How many permutations of $1, 2, 3, \ldots, 9$ are there in which exactly 5 numbers are in their original position?

7. A game uses a deck of n different cards, where n is an integer and $n \geq 6$. The number of possible sets of 6 cards that can be drawn from the deck is 6 times the number of possible sets of 3 cards that can be drawn. Find n.

(2005 AIME II)

8. Ten points are marked on a circle. How many distinct convex polygons of three or more sides can be drawn using some (or all) of the ten points as vertices?

(1989 AIME)

9. Determine the number of functions

$$f : \{1, 2, \ldots, 2014\} \to \{2015, 2016, 2017, 2018\}$$

satisfying the condition that $f(1) + f(2) + \cdots + f(2014)$ is even.

10. Compute the number of ordered pairs (x, y) of integers with $1 \leq x < y \leq 100$ such that $i^x + i^y$ is a real number where $i^2 = -1$.

11. Given a positive integer k and a set S with $|S| = n$, how many sequences $(T_1, T_2, \ldots, T_k)$ of subsets T_i of S are there such that $T_1 \subseteq T_2 \subseteq \cdots \subseteq T_k$?

12. Five regular six-sided dice are rolled. How many ways are there to roll the dice so that the total of the numbers on the five dice is 14?

13. In Awesome State, the license plates consist of three letters followed by three digits. How many possible license plates have both the letters and the digits form a palindrome? (Note: A palindrome is a sequence that reads the same backwards and forwards.)

14. 10 kids are sitting in a row. Each kid receives 1, 2, or 3 candies (all candies are identical). How many ways are there to give candy to kids so that no 2 neighbors have 4 candies in total?

15. A group of six puppies, four kittens, and three chinchillas are lined up in single file. In how many ways can they be arranged if each puppy is to be behind all smaller puppies, each kitten must stand behind all smaller kittens, and each chinchilla must sit behind all smaller chinchillas?

16. Let $(a_1, a_2, \ldots, a_{10})$ be a list of the first 10 positive integers such that for each $2 \leq i \leq 10$ either $a_i + 1$ or $a_i - 1$ or both appear somewhere before a_i in the list. How many such lists are there?

(2012 AMC 12B)

17. Let $S = \{1, 2, 3, 4, 5\}$. How many functions $f : S \to S$ satisfy

$$f(f(x)) = f(x) \text{ for all } x \in S?$$

18. In the expansion of $(2x - 3y)^7$, what is the coefficient of $x^4 y^3$?

Introductory Problems

19. Prove combinatorially that $k\binom{n}{k} = n\binom{n-1}{k-1}$.

20. Each of the students in a class writes a different 2-digit number on the whiteboard. The teacher claims that no matter what the students write, there will be at least three numbers on the whiteboard whose digits have the same sum. What is the smallest number of the students in the class for the teacher to be correct?

21. (a) Show that for any 3 integer points chosen on the line, some pair of them average to another integer.

    (b) Show that for any 5 points in $\mathbb{R}^2$ with integer coordinates, there exists some pair of them such that the midpoint of the line joining the two points also has integer coordinates.

    (c) How many points with integer coordinates do you need for the corresponding result in $\mathbb{R}^n$?

22. Suppose that $\mathcal{A}$ is a collection of subsets of $\{1, 2, \ldots, n\}$ with the property that any two sets in $\mathcal{A}$ have a non-empty intersection. Show that $\mathcal{A}$ has at most 2^{n-1} elements.

23. Prove that for $n \geq 1$,
$$\binom{n}{0} - \binom{n}{1} + \binom{n}{2} - \binom{n}{3} + \cdots + (-1)^n \binom{n}{n} = 0$$
by using the binomial theorem.

24. Show that any integer $n \geq 2$ has a prime factorization.

25. We have seen two proofs of the identity
$$2^n = \sum_{k=0}^{n} \binom{n}{k} = \binom{n}{0} + \binom{n}{1} + \cdots + \binom{n}{n},$$
one by a counting argument and one using the binomial theorem. Now prove this identity using induction on n.

26. Consider the recurrence $a_n = 28a_{n-2} - 3a_{n-1}$ with initial conditions $a_0 = 2$ and $a_1 = 19$. Find a closed form for a_n.

27. Define a graph Q_k for $k \geq 1$ to be the "k-cube graph". Each vertex of Q_k corresponds to some length k binary string. Two vertices are adjacent if and only if their strings differ in exactly 1 coordinate. How many vertices does the graph Q_k have? How many edges does Q_k have? (Note: your answers to both of these questions should be functions of k.)

28. Show that the number of vertices of odd degree in a graph must be even.

29. Give a formula for $\chi(\overline{K_n}; k)$ and $\chi(K_n; k)$. Note that $\overline{K_n}$ indicates the graph on n vertices with no edges. You may assume $k \geq \chi(K_n)$.

30. A graph $G = (V, E)$ is called *bipartite* if there exists some partition X, Y of V (i.e., $X \cup Y = V$ and $X \cap Y = \emptyset$) such that every edge of G has one endpoint in X and one endpoint in Y. Explain why every bipartite graph is 2-colorable.

31. Let b_n the number of ways to write a positive integer n as a sum of nonnegative powers of 2, where by convention we set $b_0 = 1$. Find the generating function for this sequence and use it to prove that $b_n = \sum_{k=0}^{[\frac{n}{2}]} b_k$.

32. Let n be a positive integer. Show that the number of partitions of n, where each part appears at least twice, is equal to the number of partitions of n into parts that are divisible by 2 or 3.

33. Let $f(n, k)$ be the number of ways of distributing k candies to n children so that each child receives at most 2 candies. For example $f(3, 7) = 0$, $f(3, 6) = 1$, $f(3, 4) = 6$. Determine the value of

$$f(2006, 1) + f(2006, 4) + \ldots + f(2006, 1000)$$
$$+ f(2006, 1003) + \ldots + f(2006, 4012).$$

(Adapted Canada 2006)

34. How many n digit numbers are there such that they are divisible by 3 and all their digits are either $2, 3, 7, 9$?

(Romania)

35. Consider n points $P_1, P_2, \ldots, P_n$ lying on a straight line. We color each point in white, red, green, blue and violet. A coloring is admissible if for each pair of consecutive points P_i, P_{i+1} ($i = 1, 2, \ldots, n-1$) either both points have the same color, or at least one of them is white. How many admissible collorings are there?

(Austrian-Polish 1998)

36. At a summer camp there are n girls, $G_1, G_2, \ldots, G_n$ and $2n - 1$ boys $B_1, B_2, \ldots, B_{2n-1}$. Girl G_i knows boys $B_1, B_2, \ldots, B_{2i-1}$ and no others. Prove that the number of ways to choose r boy-girl pairs so that each girl in the pair knows the boy in the pair is $\binom{n}{r}\dfrac{n!}{(n-r)!}$.

(Czech-Slovak Match 1998)

37. Let p be a positive integer, $p > 1$. Find the number of $m \times n$ tables with entries in the set $\{1, 2, \ldots, p\}$ and such that the sum of elements in each row and each column is not divisible by p.

(Mediterranean 2010)

38. Let $\mathcal{F}$ be a family of subsets of the set $\{1, 2, \ldots, n\}$ such that every element in $\mathcal{F}$ has cardinality 3 and moreover for any two distinct elements $A, B \in \mathcal{F}$ we have $|A \cap B| \leq 1$. Prove that
$$|\mathcal{F}| \leq \frac{n(n-1)}{6}.$$

39. Let S be a set of n persons such that:

    (a) Any person is acquainted to exactly k other persons in S;
    (b) Any two persons that are acquainted have exactly l common acquaintances in S;
    (c) Any two persons that are not acquainted have exactly m common acquaintances in S.

    Prove that $m(n-k) - k(k-l) + k - m = 0$.

40. A school has n students, and each student can take any number of classes. Every class has at least two students in it. We know that if two different classes have at least two common students, then the number of students in these two classes is different. Prove that the number of classes is not greater that $(n-1)^2$.

(Iran 2010)

41. There are 10001 students at an university. Some students join together to form several clubs (a student may belong to different clubs). Some clubs join together to form several societies (a club may belong to different societies). There are a total of k societies. Suppose that the following conditions hold:

    (a) Each pair of students are in exactly one club.
    (b) For each student and each society, the student is in exactly one club of the society.
    (c) Each club has an odd number of students. In addition, a club with $2m+1$ students (m is a positive integer) is in exactly m societies.

    Find all possible values of k.

(IMO Shortlist 2004)

42. Each member of a club has at most three enemies in the club. (Here enemies are mutual.) Show that the members can be divided into two groups so that each member in each group has at most one enemy in the group.

43. Several positive integers are written on a blackboard. One can erase any two distinct integers and write their greatest common divisor and least common multiple instead. Prove that eventually the numbers will stop changing.

(St Petersburg, 1996)

44. Which single squares can be removed from a 7×7 board so that the rest can be tiled with 1×3 trominos?

45. An $m \times n$ array of real numbers is given. When the sum of the numbers in any row or column is negative, we may switch the signs of all the numbers in that row or column. If this operation is iterated, prove that all of the row or column sums eventually become nonnegative.

(Russia)

46. For a $n \times n$ table with positive integer entries we are allowed to make the following operations: multiply each entry of a row by 2 or subtract 1 from each number in a column. Prove that we can always reach a table where are all the entries are 0.

47. Alfred and Bonnie play a game in which they take turns tossing a fair coin. The winner is the first person to obtain a head. They play this game several times, with the stipulation that the loser of a game goes first in the next game. Suppose Alfred goes first in the first game, what is the probability that he wins the 6th game?

(AIME 1993)

48. Let $A_1, A_2, \ldots, A_k$ be subsets of $\{1, 2, \ldots, n\}$, each with three elements. Show that it is possible to color the elements of $\{1, 2, \ldots, n\}$ with c colors, such that at most $\dfrac{k}{c^2}$ of the A_i's are monochromatic.

49. Consider a set S with n elements. Let $A_1, A_2, \ldots, A_{n+1}$ be non-empty distinct subsets of S. Then

$$\sum_{1 \leq i < j \leq n} \frac{|A_i \cap A_j|}{|A_i| \cdot |A_j|} \geq 1.$$

50. Let a_j, b_j, c_j be integers for $1 \leq j \leq N$. Assume for each j, at least one of a_j, b_j, c_j is odd. Show that there exists integers r, s, t such that $ra_j + sb_j + tc_j$ is odd for at least $\dfrac{4N}{7}$ of the values of j, $1 \leq j \leq N$.

(Putnam 2000)

51. What is the probability that n random points on a circle are contained in a semicircle?

52. A broken line with length greater than 1000 lies inside a unit square. Prove that there exists a line which intersects the broken line in at least 501 points.

53. Prove that 2015 points, no three collinear, in the plane determine at least 403 convex quadrilaterals, which have disjoint interiors.

54. Given $2n$ distinct points, no three collinear, in the plane, if we color n of them with red and n of them with blue, prove that we can connect each blue with a red point such that the pairwise segments are nonintersecting.

55. Suppose that a set S in the plane containing n points has the property that any three points can be covered by an infinite strip of width 1. Prove that S can be covered by a strip of width 2

(Balkan Math Olympiad 2010)

56. Let S be a set of points, no three collinear, with at least three points, such that for any distinct $A, B, C \in S$ the circumcenter of $\triangle ABC$ is in S. Prove that S is infinite.

Chapter 16

Advanced Problems

1. Determine the number of 8×8 matrices in which each entry is a 0 or a 1 and each row and each column contains an odd number of 1's.

2. How many seven-digit integers have exactly three distinct digits?

3. An animal shelter has n cats and $3n$ dogs. Every cat hates exactly three dogs, and no two cats hate the same dog. Find a formula for the number of ways to assign each cat a dog kennel-mate that the cat doesn't hate.

4. There are two distinguishable flagpoles, and there are 19 flags of which 10 are identical blue flags and 9 are identical green flags. Let N be the number of distinguishable arrangements using all of the flags in which each flagpole has at least one flag and no two green flags on either pole are adjacent. Find the remainder when N is divided by 1000.

$$\text{(2008 AIME II)}$$

5. Derive a closed form expression (without summations) for the number of lists of m 1s and n 0s that have k runs of 1s, where a *run* is a maximal consecutive string of identical values.

6. At a conference for superheroes and supervillains, 5 pairs of heroes and villains are giving a panel where they will sit in a row. Of course, if any superhero sits next to his or her archnemesis, complete chaos will break out and ruin the convention. How many ways can you arrange the panelists so that the program proceeds smoothly?

7. Consider 3 sets X, Y, Z with $|X| = n, |Y| = m, |Z| = r$, and $Z \subset Y$. Denote by $s_{m,n,r}$ the number of functions $f : X \to Y$ for which $Z \subseteq f(X)$. Prove that:

$$s_{m,n,r} = m^n - \binom{r}{1}(m-1)^n + \binom{r}{2}(m-2)^n - \cdots + (-1)^r (m-r)^n.$$

8. A collection of letters consists of n X's and r Y's. Find the number of different words (sequences) that can be formed from the X's and Y's if each sequence must contain n X's (and not necessarily all Y's).

9. Define an ordered triple (A, B, C) of sets to be *minimally intersecting* if $|A \cap B| = |B \cap C| = |C \cap A| = 1$ and $A \cap B \cap C = \emptyset$. For example, $(\{1, 2\}, \{2, 3\}, \{1, 3, 4\})$ is a minimally intersecting triple. Find the number of minimally intersecting ordered triples of sets for which each set is a subset of $\{1, 2, 3, 4, 5, 6, 7\}$.

(variation of 2010 AIME I # 7)

10. A *palindrome* on the alphabet $\{H, T\}$ is a sequence of H's and T's which reads the same from left to right as from right to left. Thus, $HTH, HTTH, HTHTH$, and $HTHHTH$ are palindromes of lengths $3, 4, 5$, and 6 respectively. Let $P(n)$ denote the number of palindromes of length n on $\{H, T\}$. For how many values of n is $1000 < P(n) < 10000$?

11. How many nonnegative integral solutions does the equation $n_1 + n_2 + \cdots + n_k \leq m$ have?

12. How many nonnegative integer solutions are there to $x + y + z = 30$ such that none of x, y, z is divisible by 3?

13. Find the number of triples (a, b, c) of positive integers such that each positive integer is a number from 1 to 9 and the product abc is divisible by 10.

14. Find the number of ways of arranging the numbers $1, 2, \ldots, 8$ into three non-empty sets. For example, $\{1, 3, 6, 7\}, \{2, 5\}$, and $\{4, 8\}$ is one arrangement. The order of the sets does not matter.

15. Let S_1 and S_2 represent two binary strings of length n. The *Hamming distance* between S_1 and S_2, denoted by $\mathcal{H}(S_1, S_2)$, is the number of positions in which S_1 and S_2 differ. For example, $\mathcal{H}(001011, 101001) = 2$. Given positive integers n and k with $k \leq n$, count the number of ordered pairs (S_1, S_2) of two binary strings S_1 and S_2, each of length n, such that $\mathcal{H}(S_1, S_2) = k$.

16. How many 15-letter arrangements of 5 A's, 5 B's, and 5 C's have no A's in the first 5 letters, no B's in the next 5 letters, and no C's in the last 5 letters?

(variation on 2003 AMC 12A)

17. Consider numbers such that all digits of the numebr are different, the first digit is not zero, and the sum of the digits is 36. There are $N \times 7!$ such numbers. What is the value of N?

18. In a sequence of coin tosses one can keep a record of instances in which a tail is immediately followed by a head, a head is immediately followed by a head, etc. We denote these by TH, HH, etc. For example, in the sequence $HHTTHHHHTHHTTTT$ of 15 coin tosses we observe that there are five HH, three HT, two TH, and four TT subsequences. How many different sequences of 15 coin tosses will contain exactly two HH, three HT, four TH, and five TT subsequences?

(1986 AIME)

19. The expression
$$(x+y+z)^{2006} + (x-y-z)^{2006}$$
is simplified by expanding it and combining like terms. How many terms are in the simplified expression?

(2006 AMC 12A)

20. The polynomial $1 - x + x^2 - x^3 + \cdots + x^{16} - x^{17}$ may be written in the form $a_0 + a_1 y + a_2 y^2 + \cdots + a_{16} y^{16} + a_{17} y^{17}$, where $y = x + 1$ and the a_i's are constants. Find the value of a_2.

(1986 AIME)

21. Let S be a set containing n elements. Prove that
$$\sum_{A \subseteq S} \sum_{B \subseteq S} |A \cap B| = n \cdot 4^{n-1}.$$

22. Show that any odd number not divisible by 5 must divide some number of the form $10101\ldots01$, an alternating string of 1's and 0's. For example, 13 divides 10101, 17 divides 101010101010101, 9 and 19 divide 10101010101010101.

23. Show that for every 16-digit number there is a string of one or more consecutive digits such that the product of these digits is a perfect square.

(1991 Japan Mathematical Olympiad)

24. 10 integers are chosen from 1 to 100. Prove that we can find 2 disjoint, non-empty subsets of the chosen integers such that the 2 subsets give the same sum of elements.

25. Every point in a plane is either red, green, or blue. Prove that there exists a rectangle in the plane such that all of its vertices are the same color.

(USAMTS Year 18)

26. Jenny starts with a pile of n stones, where $n \geq 2$ is a positive integer. At each step, she takes a pile of stones and splits it into two smaller piles. If the two new piles have a stones and b stones, then she writes the product ab on a blackboard. She keeps repeating these steps, until each pile has exactly one stone. Prove that no matter how Jenny splits the stones, the sum of the numbers on the blackboard at the end is always the same.

(For example, if Jenny starts with a pile of 12 stones, she can split it into a pile of 5 stones and a pile of 7 stones, and writes the number $5 \cdot 7 = 35$ on the blackboard. She can then split the pile of 5 stones into a pile of 2 stones and a pile of 3 stones, and writes the number $2 \cdot 3 = 6$ on the blackboard.)

27. Prove that $\chi(T; k) = k(k-1)^{n-1}$ for all trees T on n vertices.

28. Show that every graph on n vertices with at least n edges contains a cycle.

29. Let $p > 2$ be a prime number.
Find the number of subsets of $\{1, 2, \ldots, p-1\}$ with sum divisible by p.

(Bulgaria, 2006)

30. A deck of 32 cards has 2 different jokers each of which is numbered 0. There are 10 red cards numbered 1 through 10 and similarly for blue and green cards. One chooses a number of cards from the deck to form a hand. If a card in the hand is numbered k, then the value of the card is 2^k, and the value of the hand is sum of the values of the cards in hand. Determine the number of hands having the value 2004.

(CGMO 2004)

31. The set of natural numbers is partitioned into finitely many arithmetic progressions $\{a_i + dr_i\}$, $1 \leq i \leq n$. Prove that:

(a) $\sum_{i=1}^{n} \dfrac{1}{r_i} = 1.$

(b) There exist $i \neq j$ such that $r_i = r_j$.

(c) $\sum_{i=1}^{n} \dfrac{a_i}{r_i} = \dfrac{n-1}{2}.$

Advanced Problems

32. Find all natural numbers n for which there exist two distinct sets of integers $\{a_1, a_2, \ldots, a_n\}$ and $\{b_1, b_2, \ldots, b_n\}$ such that the multisets

$$\{a_i + a_j | 1 \leq i < j \leq n\} \quad \text{and} \quad \{b_i + b_j | 1 \leq i < j \leq n\},$$

coincide.

(Erdös-Selfridge)

33. Determine with proof whether there exist a subset X of the nonnegative integers with the following property : for any integer n there is exactly one solution to $a + 2b = n$ with $a, b \in X$.

(Adapted from USAMO 1996)

34. Let n be a positive integer, and $X = \{1, 2, \ldots, 2n\}$. How many subsets S of X are there, such that no two elements $x, y \in S$ differ by 2?

35. A rectangle is divided into 200×3 unit squares. Prove that the number of ways of splitting this rectangle into rectangles of size 1×2 is divisible by 3.

(Baltic Way 2005)

36. A word is a sequence of n letters of the alphabet $\{a, b, c, d\}$. A word is said to be complicated if it contains two consecutive groups of identic letters. The words *caab*, *baba* and *cababdc*, for example, are complicated words, while *bacba* and *dcbdc* are not. A word that is not complicated is a simple word. Prove that the numbers of simple words with n letters is greater than 2^n, if n is a positive integer.

(Romania TST 2003)

37. A permutation $\sigma : \{1, 2, \ldots, n\} \to \{1, 2, \ldots, n\}$ is called straight if and only if for each integer k, $1 \leq k \leq n-1$ the following inequality is fulfilled

$$|\sigma(k) - \sigma(k-1)| \leq 2.$$

Find the smallest integer n for which there exist at least 2003 straight permutations.

(Romania TST 2003)

38. 16 students took part in a mathematical competition where every problem was a multiple choice question with four choices. After the contest, it is found that any two students had at most one answer in common. Prove that there are at most 5 problems in the contest.

(China 1992)

39. A group of 10 people went to a bookstore. It is known that everyone bought exactly 3 books and for every two persons, there is at least one book both of them bought. What is the least number of people that could have bought the book purchased by the greatest number of people?

(China 1993)

40. Let X be a set with n elements. Given $k > 2$ subsets of X, each with at least r elements, show that we can find two of them whose intersection has at least $r - \dfrac{nk}{4(k-1)}$ elements.

(Iberoamerican 2001)

41. Let X be a finite set with n elements and let $A_1, A_2, \ldots, A_m$ be three element subsets of X such that $|A_i \cap A_j| \leq 1$ for all $i \neq j$. Show that there exists a subset A of X with at least $\lfloor \sqrt{2n} \rfloor$ elements containing none of the A_i's.

42. Let T be a finite set of integers greater than 1. A subset S of T is *good* if for any $t \in T$ one can find $s \in S$ such that s and t are not relatively prime. Prove that the number of *good* subsets of T is odd.

(USA TST 2010)

43. Prove that for any set of n points in the plane there at most $cn\sqrt{n}$ distances among these points equal to 1, for some absolute constant $c > 0$.

44. The numbers from 1 through 2015 are written on a blackboard. Every second, Dr. Math erases four numbers of the form a, b, c, $a+b+c$, and replaces them with the numbers $a+b$, $b+c$, $c+a$. Prove that this can continue for at most 9 minutes.

45. Several stones are placed on an infinite (in both directions) strip of squares, indexed by the integers. We perform a sequence of moves, each move being one of the following two types:

(a) Remove one stone from each of the squares $n-1$ and n and place one stone on square $n+1$.

(b) Remove two stones from square n and place one stone on each of the squares $n-2$ and $n+1$.

Prove that any sequence of such moves will lead to a position in which no further moves can be made, and moreover that this position is independent of the sequence of moves.

(Russia 1997)

46. Consider a matrix whose entries are integers. Adding the same integer to all entries on a row, or in a column, is called an operation. It is given that, for infinitely many positive integers n, one can obtain, through a finite number of operations, a matrix having all entries divisible by n. Prove that, through a finite number of operations, one can obtain the null matrix.

(Romania 2009)

47. At the vertices of a regular hexagon six nonnegative integers are written whose sum is n. One is allowed to make moves of the following form: (s)he may pick a vertex and replace the number written there by the absolute value of the difference between the numbers written at the two neighboring vertices. Prove that if n is odd then one can make a sequence of moves, after which the number 0 appears at all six vertices.

(USAMO 2003)

48. A $(2n+1) \times (2n+1)$ board is going to be tiled with pieces of the types as shown

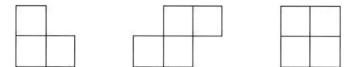

where rotations and reflections of the tiles are allowed.

Prove that at least $4n + 3$ pieces of the first type will be used.

(IMO Shortlist 2002)

49. A regular 2004-sided polygon is given, with all of its diagonals drawn. After some sides and diagonals are removed, every vertex has at most five segments coming out of it. Prove that one can color the vertices with two colors such that at least $\dfrac{3}{5}$ of the remaining segments have ends with different colors.

(MOP Homework 2004)

50. Let A be a set of N residues modulo N^2. Prove that there exists a set B of N residues modulo N^2 such that $A + B = \{a+b | a \in A, b \in B\}$ has at least $\dfrac{N^2}{2}$ elements.

(IMO Shortlist 1999)

51. An $m \times n$ checkerboard is colored randomly: each square is independently assigned red or black with probability $\frac{1}{2}$. we say that two squares, p and q, are in the same connected monochromatic region if there is a sequence of squares, all of the same color, starting at p and ending at q, in which successive squares in the sequence share a common side. Show that the expected number of connected monochromatic regions is greater than $\frac{mn}{8}$.

(Putnam 2004)

52. A finite collection of squares has total area 4. Show that they can be arranged to cover a square of side 1.

53. Given $2n + 3$ points in the plane, no three collinear and no four on a circle, prove that there exists a circle containing three of the points such that exactly n of the remaining points are in its interior.

54. Let $n \geq 4$ be a fixed positive integer. Given a set $S = \{P_1, P_2, \ldots, P_n\}$ of n points in the plane such that no three are collinear and no four concyclic, let a_t, $1 \leq t \leq n$, be the number of circles $P_i P_j P_k$ that contain P_t in their interior, and let $m(S) = \sum_{i=1}^{n} a_i$. Prove that there exists a positive integer $f(n)$, depending only on n, such that the points of S are the vertices of a convex polygon if and only if $m(S) = f(n)$.

(IMO Shortlist 2000)

55. Let S be a set of n points in the plane. No three points of S are collinear. Prove that there exists a set P containing $2n - 5$ points satisfying the following condition: In the interior of every triangle whose three vertices are elements of S lies a point that is an element of P.

(IMO Shortlist 1991)

56. $\mathcal{A}$ is a closed polygon set so that for any two points in $\mathcal{A}$, the line segment joining the two points lies completely in $\mathcal{A}$. Prove that there exists a point O in $\mathcal{A}$, such that for any points X, X' on the boundary of $\mathcal{A}$, such that O lies on line segment XX' we have

$$\frac{1}{2} \leq \frac{OX}{OX'} \leq 2.$$

(Iran 2004)

Chapter 17

Solutions for Introductory Problems

1. Let k be a positive integer. In how many ways can one select a subset of three distinct numbers from the set $\{1, 2, \ldots, 3k\}$ such that their sum is divisible by 3?

   Solution. In order for the sum of three numbers to be divisible by 3, either all of the chosen numbers must have the same remainder upon division by 3 or one has remainder 0, one has remainder 1, and the last has remainder 2. We will consider each case separately.

   Suppose all three of our chosen numbers are multiples of 3 (i.e., they all have remainder 0 upon division by 3). There are k such numbers in the set $\{1, 2, \ldots, 3k\}$ (namely $\{3, 6, 9, \ldots, 3k\}$) and we wish to choose 3 of them. The number of ways to do this is $\binom{k}{3}$. Similarly, since there are k numbers in our set with remainder 1 upon division by 3, there are $\binom{k}{3}$ ways for all three of our numbers to have remainder 1. The same holds for remainder 2. Thus, there are $\binom{k}{3} + \binom{k}{3} + \binom{k}{3} = 3\binom{k}{3}$ ways for all three of our numbers to have the same remainder upon division by 3.

   Now suppose all three of our numbers have different remainders upon division by 3. There are k choices for which number with remainder 0 we choose, k choices for which number with remainder 1 we choose, and k choices for which number with remainder 2 we choose for a total of $k \cdot k \cdot k = k^3$ ways to have a set of three different numbers from our set with different remainders upon division by 3.

   Adding our results from these two cases up, we find there are $3\binom{k}{3} + k^3$ ways to select our subset such that the sum of the elements is divisible by 3. □

2. How many n-digit numbers have their digits in non-decreasing order (i.e., each digit is greater than or equal to each digit to its right)? For example, the number 122379999 has its digits in non-decreasing order, but the numbers 12330 and 13572468 do not.

   Solution. First, note that any number with its digits in non-decreasing order cannot have 0 as a digit. Notice also that once we have chosen our n digits, the order in which they must appear in order to form a number with its digits in non-decreasing order is fixed. Thus, this problem is really asking how many ways there are to choose n digits from the set $\{1, 2, \ldots, 9\}$ allowing for repetition. We can solve this problem using stars and bars where the n stars represent a particular occurrence of a digit in our number and the 8 bars divide the stars into different digits $1, 2, \ldots, 9$. The number of arrangements of n stars and 8 bars is $\binom{n+8}{n}$ so this is how many n-digit numbers have their digits in non-decreasing order. □

3. How many North-East lattice paths are there from $(0,0)$ to $(8,8)$ that never touch the point $(4,6)$ or the point $(2,3)$?

   Solution. We know the total number of paths from $(0,0)$ to $(8,8)$ is $\binom{16}{8}$ since we must take 16 total steps and we need to choose 8 of those to be to the right. Now we subtract off those paths from $(0,0)$ to $(8,8)$ which pass through the point $(4,6)$. There are $\binom{10}{4}$ paths from $(0,0)$ to $(4,6)$, and there are $\binom{6}{4}$ paths from $(4,6)$ to $(8,8)$. Thus overall there are $\binom{10}{4} \cdot \binom{6}{4}$ paths from $(0,0)$ to $(8,8)$ which touch $(4,6)$. Similarly, there are $\binom{5}{2} \cdot \binom{11}{6}$ paths from $(0,0)$ to $(8,8)$ which touch $(2,3)$. We then must add back those paths that pass through both $(4,6)$ and $(2,3)$. There are $\binom{5}{2}$ paths from $(0,0)$ to $(2,3)$, $\binom{5}{2}$ paths from $(2,3)$ to $(4,6)$, and $\binom{6}{4}$ paths from $(4,6)$ to $(8,8)$. Thus overall we have

   $$\binom{16}{8} - \binom{10}{4} \cdot \binom{6}{4} - \binom{5}{2} \cdot \binom{11}{6} + \binom{5}{2} \cdot \binom{5}{2} \cdot \binom{6}{4} = 6600$$

   total paths from $(0,0)$ to $(8,8)$ which never touch $(4,6)$ or $(2,3)$. □

4. A committee of 5 is to be chosen from a group of 10 people. How many ways can it be chosen if Dave and Richard must serve together or not at all, and Tina and Val refuse to serve with each other?

   Solution. We look at two cases: the case where Dave and Richard are on the committee together and the case where neither of them is. If both are on the committee, we must choose the remaining 3 members of the committee from the remaining 8 people. There are $\binom{8}{3}$ ways to choose 3 of 8 people, but this counts the possibility that Tina and Val are both on the committee, so we must subtract those cases off. If Dave, Richard, Tina, and Val are all on the committee, we just have 1 more person to choose from the remaining 6 so there are $\binom{6}{1}$ ways to do this.

On the other hand, if neither Dave nor Richard are on the committee, we must pick 5 of the remaining 8 people for the committee. This can be done in $\binom{8}{5}$ ways. Again, however, we have included committees with both Tina and Val. These committees will have Tina, Val, and 3 of the remaining 6 people, so there are $\binom{6}{3}$ ways to form such a committee. This gives a total of

$$\binom{8}{3} - \binom{6}{1} + \binom{8}{5} - \binom{6}{3} = 56 - 6 + 56 - 20 = 86$$

committees that satisfy the given conditions. $\square$

5. Call a number *prime-looking* if it is composite but not divisible by 2, 3, or 5. The three smallest prime-looking numbers are 49, 77, and 91. There are 168 prime numbers less than 1000. How many prime-looking numbers are there less than 1000?

(2005 AMC 12A)

Solution. We use the Principle of Inclusion-Exclusion to count how many numbers less than or equal to 1000 are multiples of 2, 3, or 5. We know the number of multiples of 2 is $1000/2 = 500$, the number of multiples of 3 is $\lfloor 1000/3 \rfloor = 333$, and the number of multiples of 5 is $1000/5 = 200$. However we must subtract off the numbers that are multiples of any pair of these numbers. There are $\lfloor 1000/6 \rfloor = 166$ multiples of $2 \cdot 3 = 6$, $1000/10 = 100$ multiples of $2 \cdot 5 = 10$, and $\lfloor 1000/15 \rfloor = 66$ multiples of $3 \cdot 5 = 15$. Finally we need to add back those numbers that are multiples of all three. There are $\lfloor 1000/30 \rfloor = 33$ multiples of $2 \cdot 3 \cdot 5 = 30$. This gives us $500 + 333 + 200 - 166 - 100 - 66 + 33 = 734$ numbers less than or equal to 1000 that are multiples of either 2, 3, or 5.

Since we want to know how many composite numbers less than 1000 are not divisible by 2, 3, or 5, we find

$$1000 - 734 - 168 + 3 - 1 = 100$$

prime-looking numbers less than 1000. Note that we must add back 3 because 2, 3, and 5 are both multiples of 2, 3, or 5 and prime. We subtract off 1 because 1 is neither prime nor composite. $\square$

6. How many permutations of $1, 2, 3, \ldots, 9$ are there in which exactly 5 numbers are in their original position?

   Solution. We first choose which 5 numbers are in their original positions. There are $\binom{9}{5}$ ways to do this. Once we have fixed these 5 numbers, we must derange the remaining 4 (see Example 34). We know the number of ways to derange 4 objects is

   $$4! - \frac{4!}{1!} + \frac{4!}{2!} - \frac{4!}{3!} + \frac{4!}{4!} = 9$$

   so there are $\binom{9}{5} \cdot 9 = 1134$ such permutations. □

7. A game uses a deck of n different cards, where n is an integer and $n \geq 6$. The number of possible sets of 6 cards that can be drawn from the deck is 6 times the number of possible sets of 3 cards that can be drawn. Find n.

   (2005 AIME II)

   Solution. The number of ways to choose a set of 6 cards from a deck of n total cards is $\binom{n}{6}$, and the number of ways to choose a set of 3 cards from a deck of n total cards is $\binom{n}{3}$. Thus we have $\binom{n}{6} = 6\binom{n}{3}$. Substituting in our formula for the binomial coefficients, we have

   $$\frac{n!}{(n-6)!6!} = 6\frac{n!}{(n-3)!3!}$$

   Canceling out an $n!$ from each side and cross multiplying, we have

   $$(n-3)!3! = 6(n-6)!6! \Rightarrow (n-3)(n-4)(n-5) = 720.$$

   Thus our goal is to find three consecutive integers whose product is 720. Since $720 = 8 \cdot 9 \cdot 10$, we know $n - 3 = 10$ and thus $n = 13$. □

8. Ten points are marked on a circle. How many distinct convex polygons of three or more sides can be drawn using some (or all) of the ten points as vertices?

   (1989 AIME)

   Solution. A convex polygon with k sides is uniquely formed by choosing k of the points on the circle and drawing the chords between consecutive

Solutions for Introductory Problems 167

pairs of points clockwise around the circle. There are $\binom{10}{k}$ ways to choose k of the 10 points, so the number of convex polygons is

$$\sum_{k=3}^{10} \binom{10}{k} = \sum_{k=0}^{10} \binom{10}{k} - \binom{10}{2} - \binom{10}{1} - \binom{10}{0}$$

$$= 2^{10} - 45 - 10 - 1 = 968. \qquad \square$$

9. Determine the number of functions

$$f : \{1, 2, \ldots, 2014\} \to \{2015, 2016, 2017, 2018\}$$

satisfying the condition that $f(1) + f(2) + \cdots + f(2014)$ is even.

Solution. There are 4 choices for each $f(i)$. This means there are 4^{2013} ways to assign values to $f(1), f(2), \ldots, f(2013)$. At that point we consider $s = f(1) + f(2) + \cdots + f(2013)$. If s is odd, $f(2014)$ must also be odd in order for the total sum to be even. This implies $f(2014)$ is either 2015 or 2017. On the other hand, if s is even, $f(2014)$ must also be even, so again we have two choices for $f(2014)$ (2016 or 2018). Thus there are $2 \cdot 4^{2013}$ functions f such that $f(1) + f(2) + \cdots + f(2014)$ is even. $\square$

10. Compute the number of ordered pairs (x, y) of integers with $1 \leq x < y \leq 100$ such that $i^x + i^y$ is a real number where $i^2 = -1$.

Solution. In order for $i^x + i^y$ to be a real number, either both i^x and i^y must themselves be real numbers or we must have $i^x + i^y = 0$ with i^x and i^y imaginary. We will examine each of these cases in turn. If i^x and i^y are both real, x and y must both be even numbers. Since there are 50 even numbers between 1 and 100 inclusive, there are $\binom{50}{2}$ ways to select x and y (since $x < y$ once we choose two numbers we know which must be x and which must be y). Alternatively, we could have $i^x = -i^y$ with both imaginary. This will occur when one of the numbers has a remainder of 1 upon division by 4 and the other has a remainder of 3 upon division by 4. There are 25 of each of these from 1 to 100 inclusive, so there are $25 \cdot 25 = 25^2$ ways to choose x and y in this case (as before, once the two numbers are chosen it is predetermined which is x and which is y). Thus, there are $\binom{50}{2} + 25^2 = 1850$ total ordered pairs (x, y) such that $1 \leq x < y \leq 100$ and $i^x + i^y$ is a real number. $\square$

11. Given a positive integer k and a set S with $|S| = n$, how many sequences $(T_1, T_2, \ldots, T_k)$ of subsets T_i of S are there such that $T_1 \subseteq T_2 \subseteq \cdots \subseteq T_k$?

Solution. We will consider each element of S one at a time. Note that if an element $x \in S$ is in T_i, then $x \in T_j$ for all $j \geq i$. Thus we can uniquely determine the subsets by deciding which subset an element first appears in: $T_1, T_2, \ldots, T_k$ or if it does not appear at all. This is a total of $k+1$ possibilities for each of the n elements of S, giving a total of $(k+1)^n$ sequences of subsets of S such that $T_1 \subseteq T_2 \subseteq \cdots \subseteq T_k$. □

12. Five regular six-sided dice are rolled. How many ways are there to roll the dice so that the total of the numbers on the five dice is 14?

    Solution. Let n_1, n_2, n_3, n_4, n_5 be the numbers rolled on each die. So we are essentially counting the positive integral solutions to $n_1 + n_2 + n_3 + n_4 + n_5 = 14$ where each n_i is at most 6. If we let $n_1 = \ell_1 + 1$ we can simplify our problem even further, looking for nonnegative integral solutions to $\ell_1 + \ell_2 + \ell_3 + \ell_4 + \ell_5 = 9$ where each ℓ_i is at most 5. We will use complementary counting: first counting without the at most 5 restriction, then subtracting off the possible solutions where at least one of the ℓ_i is greater than 5.

    We know from Example 21 that the number of nonnegative integral solutions to $\ell_1 + \ell_2 + \ell_3 + \ell_4 + \ell_5 = 9$ is $\binom{13}{4}$. Now suppose that at least one of the ℓ_i is greater than 5. Notice that since all ℓ_i are nonnegative and their sum is 9, at most one can be greater than 5 at a time. There are $\binom{5}{1} = 5$ ways to choose which of the five variables will be greater than 5. Once chosen, we can assign 6 units to this variable to ensure it is greater than 5, leaving 3 extra units to be assigned amongst all 5. Note that this allows the chosen variable to end up with a value greater than 6 so that we can avoid extra casework. The number of ways to assign these 3 units will be $\binom{7}{4}$, so our final answer is

    $$\binom{13}{4} - 5 \cdot \binom{7}{4} = 715 - 5 \cdot 35 = 540$$

    possible rolls. □

13. In Awesome State, the license plates consist of three letters followed by three digits. How many possible license plates have both the letters and the digits form a palindrome? (Note: A palindrome is a sequence that reads the same backwards and forwards.)

    Solution. We have 26 choices for the first letter and 26 choices for the second letter as well. Since the letters must form a palindrome, the third letter will be the same as the first so there are 26^2 ways to choose

Solutions for Introductory Problems 169

the three letters of the license plate. Similarly, there are 10 choices for the first digit and 10 choices for the second digit, but the third must match the first in order to form a palindrome. In total this gives us $26^2 \cdot 10^2 = 67600$ license plates such that both the letter and digits form a palindrome. $\square$

14. 10 kids are sitting in a row. Each kid receives 1, 2, or 3 candies (all candies are identical). How many ways are there to give candy to kids so that no 2 neighbors have 4 candies in total?

    Solution. We begin with the child on the left and move down the row to the right. The first kid can receive 1, 2, or 3 candies so there are 3 choices for how many candies he will receive. For the next child, there will be exactly one number among 1, 2, and 3 that adds with the first child's candies to make 4, so we have only 2 choices for how many candies this child will receive. Similarly, the next child only has 2 choices of how many candies to receive so that her candies do not sum to 4 with the second child's candies. This continues for each of the remaining 7 children to give us a total of $3 \cdot 2^9 = 1536$ possible distributions of candy. $\square$

15. A group of six puppies, four kittens, and three chinchillas are lined up in single file. In how many ways can they be arranged if each puppy is to be behind all smaller puppies, each kitten must stand behind all smaller kittens, and each chinchilla must sit behind all smaller chinchillas?

    Solution. We can think of this as the number of ways to arrange six P's, four K's, and three C's, since once we know which positions the animals of a particular species occupy, we know which order they must stand in (namely smallest to largest from front to back). We know the number of arrangements of these letters is $\binom{13}{6,4,3} = 60,060$ so there are $60,060$ lines of animals meeting the given criteria. $\square$

16. Let $(a_1, a_2, \ldots, a_{10})$ be a list of the first 10 positive integers such that for each $2 \leq i \leq 10$ either $a_i + 1$ or $a_i - 1$ or both appear somewhere before a_i in the list. How many such lists are there?

    (2012 AMC 12B)

    Solution. We will condition on the first number a_1. If $a_1 = k$, we must choose $k-1$ of the remaining 9 spots to contain the integers less than k. We know these integers must appear in the order $k-1, k-2, \ldots, 1$ to ensure that $a_i + 1$ appears before them on the list. Similarly, the integers greater than k must appear in the order $k+1, k+2, \ldots, 10$ to ensure

$a_i - 1$ appears before them on the list. Since the values in a particular slot are determined as soon as we choose which spots contain the $k-1$ integers smaller than k, there are $\binom{9}{k-1}$ lists meeting the criteria with $a_1 = k$. Summing over all values of k, we have

$$\sum_{k=1}^{10} \binom{9}{k-1} = \sum_{j=0}^{9} \binom{9}{j} = 2^9 = 512 \text{ valid lists.} \qquad \square$$

17. Let $S = \{1, 2, 3, 4, 5\}$. How many functions $f : S \to S$ satisfy

$$f(f(x)) = f(x) \text{ for all } x \in S?$$

Solution. Notice that if $s \in f(S)$ (i.e., s is in the image of f), then $f(s) = s$ since we must have $f(f(x)) = f(x)$ for all $x \in S$. We do casework based on the size of $f(S)$ (i.e., how many elements are in the image of f).

- There is only one way to have all five elements of S be in $f(S)$. In this case, $f(x) = x$ for all $x \in S$.
- If there are four elements of S in $f(S)$, there are $\binom{5}{4} = 5$ ways to choose which four elements are in $f(S)$. Each of these elements must map to itself. There are then 4 choices for what the remaining element will map to since it cannot map to itself. This gives a total of $5 \cdot 4 = 20$ functions.
- If there are three elements of S in $f(S)$, there are $\binom{5}{3} = 10$ ways to choose which three they are. There are 3 ways each to assign the mapping of the remaining two elements under f for a total of $10 \cdot 3^2 = 90$ functions.
- If there are two elements of S in $f(S)$, there are $\binom{5}{2} = 10$ ways to choose those two elements. Then for each of the other three elements we have 2 choices for what f maps them to. This gives us $10 \cdot 2^3 = 80$ different functions.
- If only one element of S is in $f(S)$, every element of S maps to that element so there are only 5 functions f (one for each element of S).

Since this covers all possible cases for the size of $f(S)$, we sum them to obtain $1 + 20 + 90 + 80 + 5 = 196$ functions from S to S satisfying $f(f(x)) = f(x)$ for all $x \in S$. $\qquad \square$

18. In the expansion of $(2x - 3y)^7$, what is the coefficient of $x^4 y^3$?

Solution. We know by the binomial theorem that

$$(x+y)^n = \sum_{k=0}^{n} \binom{n}{k} x^k y^{n-k}$$

so we know

$$(2x-3y)^7 = \sum_{k=0}^{7} \binom{7}{k} (2x)^k (-3y)^{n-k}$$

This tells us the $x^4 y^3$ term will be

$$\binom{7}{4}(2x)^4(-3y)^3 = -15120 x^4 y^3.$$

Thus the coefficient is -15120. $\square$

19. Prove combinatorially that $k\binom{n}{k} = n\binom{n-1}{k-1}$.

Solution. We use a committee forming argument here, counting the number of ways to pick a committee of k people, one of whom is the chairman, from a class of n students.

We know there are $\binom{n}{k}$ ways to choose a group of k students from n total students. Once we have these k people, there are then $\binom{k}{1} = k$ ways to choose who will be the chairperson. This gives a total of $k\binom{n}{k}$ ways to choose our committee.

Alternatively, we could first choose our chairman from the n students in the class. There are $\binom{n}{1} = n$ ways to do this. We then pick the other $k-1$ committee members from the remaining $n-1$ students; this can be done in $\binom{n-1}{k-1}$ ways for a total of $n\binom{n-1}{k-1}$ committees.

Since these two quantities count the same thing, they must be equal. Thus we have $k\binom{n}{k} = n\binom{n-1}{k-1}$ as desired. $\square$

20. Each of the students in a class writes a different 2-digit number on the whiteboard. The teacher claims that no matter what the students write, there will be at least three numbers on the whiteboard whose digits have the same sum. What is the smallest number of the students in the class for the teacher to be correct?

Solution. We solve this problem by applying the Pigeonhole Principle. Let N be the number of students. At most one student can write the number 10 on the board and at most one student can write the number

99. Break the remaining two digit numbers into sets A_k with digit sum equal to k. Thus $A_2 = \{11, 20\}$, $A_3 = \{12, 21, 30\}$, ..., $A_{17} = \{89, 98\}$. There are 16 such sets and these will be our "holes". The remaining $N - 2$ students will be our pigeons. If $N - 2 > 32$, then there must be one of these 16 holes that gets 3 students, and hence the teacher will be correct. Thus the teacher is correct for classes with $N \geq 35$ students.

If $N = 34$, then it is easy to see that the teacher can be wrong. For example, if the students write down the 34 numbers 10, 11, 12,..., 29, 38, 39, 48, 49,..., 98, 99, there will be no three numbers on the whiteboard whose digits have the same sum, since the numbers chosen are 11, 99, and the two smallest numbers in each of the sets A_k.

Thus, the smallest number of students in the class for the teacher to be correct is 35. □

21. (a) Show that for any 3 integer points chosen on the line, some pair of them average to another integer.

  (b) Show that for any 5 points in $\mathbb{R}^2$ with integer coordinates, there exists some pair of them such that the midpoint of the line joining the two points also has integer coordinates.

  (c) How many points with integer coordinates do you need for the corresponding result in $\mathbb{R}^n$?

Solution.

  (a) If we have 3 integers, by the Pigeonhole Principle at least two of these numbers must have the same parity (odd or even). Since the sum of two numbers with the same parity is even and every even number is a multiple of 2, the average of our pair of numbers with the same parity will be an integer.

  (b) Suppose we have 5 points with integer coordinates in $\mathbb{R}^2$. There are $2 \cdot 2 = 4$ possible parity combinations for the two coordinates: (odd, odd), (odd, even), (even, odd), and (even, even). If we have 5 points, by the Pigeonhole Principle we know at least two of them must have the same parity combination. Recall that the coordinates of the midpoint of the line segment between two points are the average of the corresponding coordinates of the two endpoints. Thus if we take the midpoint of these two points, it will have integer coordinates based on our result in part a.

  (c) The result in $\mathbb{R}^n$ is: for any $2^n + 1$ points in $\mathbb{R}^n$ with integer coordinates, there exists some pair of them such that the midpoint of the line joining the two points also has integer coordinates. This is due to the fact that there are 2^n possible parity combinations for points in $\mathbb{R}^n$. $\square$

22. Suppose that $\mathcal{A}$ is a collection of subsets of $\{1, 2, \ldots, n\}$ with the property that any two sets in $\mathcal{A}$ have a non-empty intersection. Show that $\mathcal{A}$ has at most 2^{n-1} elements.

    Solution. We know that there are 2^n total subsets of the set $\{1, 2, \ldots, n\}$. We will create 2^{n-1} pairs of subsets with each pair consisting of a subset of $\{1, 2, \ldots, n\}$ and its complement. In order for $\mathcal{A}$ to satisfy the property that any two sets in $\mathcal{A}$ have a non-empty intersection, we cannot have both a set and its complement contained in $\mathcal{A}$. However, if $\mathcal{A}$ has more than 2^{n-1} elements, by the Pigeonhole Principle there must be one of our pairs such that both subsets are in $\mathcal{A}$. But this violates our property, so $\mathcal{A}$ must have at most 2^{n-1} elements in order for our property to hold. $\square$

23. Prove that for $n \geq 1$,
    $$\binom{n}{0} - \binom{n}{1} + \binom{n}{2} - \binom{n}{3} + \cdots + (-1)^n \binom{n}{n} = 0$$
    by using the binomial theorem.

    Solution. The binomial theorem tells us that
    $$(x+y)^n = \binom{n}{0} x^0 y^n + \binom{n}{1} x^1 y^{n-1} + \cdots + \binom{n}{n} x^n y^0.$$
    Setting $x = -1, y = 1$, this yields
    $$0 = (-1+1)^n$$
    $$= \binom{n}{0}(-1)^0 1^n + \binom{n}{1}(-1)^1 1^{n-1} + \cdots + \binom{n}{n}(-1)^n 1^0$$
    $$= \binom{n}{0} - \binom{n}{1} + \binom{n}{2} - \binom{n}{3} + \cdots + (-1)^n \binom{n}{n}$$
    as desired. $\square$

24. Show that any integer $n \geq 2$ has a prime factorization.

    Solution. We proceed by strong induction on n.

    <u>Base Case</u>: ($n = 2$) 2 is prime, so it is its own prime factorization.

    <u>Induction Hypothesis</u>: Suppose n has a prime factorization for all $2 \leq n \leq k$.

Inductive Step: Consider $k+1$. There are two possibilities since $k+1$ is an integer greater than 2: either $k+1$ is prime or it is composite. If $k+1$ is prime, it is its own prime factorization and we are done. Otherwise, there exist some integers a, b with $1 < a, b < k+1$ such that $k+1 = ab$. By our strong induction hypothesis, this implies a and b both have prime factorizations. The product of these prime factorizations gives us in turn a prime factorization of $k+1$.

By the principle of mathematical induction, this concludes our proof. □

25. We have seen two proofs of the identity

$$2^n = \sum_{k=0}^{n} \binom{n}{k} = \binom{n}{0} + \binom{n}{1} + \cdots + \binom{n}{n},$$

one by a counting argument and one using the binomial theorem. Now prove this identity using induction on n.

Solution. Base Case: ($n = 0$), We have $2^0 = 1 = \binom{0}{0}$, so our identity holds when $n = 0$.

Induction Hypothesis: Suppose that for some $m \geq 0$, we know

$$2^m = \sum_{k=0}^{m} \binom{m}{k}.$$

Inductive Step: Consider 2^{m+1}. We know this is $2 \cdot 2^m$, so by our induction hypothesis we can substitute and rearrange to obtain

$$2^{m+1} = 2 \sum_{k=0}^{m} \binom{m}{k} = \binom{m}{0} + \binom{m}{m} + \sum_{k=1}^{m} \left(\binom{m}{k} + \binom{m}{k-1} \right).$$

Applying Pascal's Identity and taking advantage of the fact that

$$\binom{m}{0} = \binom{m+1}{0} = \binom{m}{m} = \binom{m+1}{m+1} = 1,$$

we have

$$2^{m+1} = \binom{m}{0} + \binom{m}{m} + \sum_{k=1}^{m} \binom{m+1}{k} = \sum_{k=0}^{m+1} \binom{m+1}{k}$$

as desired. By the principle of mathematical induction, this concludes our proof. □

Solutions for Introductory Problems

26. Consider the recurrence $a_n = 28a_{n-2} - 3a_{n-1}$ with initial conditions $a_0 = 2$ and $a_1 = 19$. Find a closed form for a_n.

    Solution. Suppose our solution has the form $a_n = c \cdot r^n$ for some constants c and r. We can substitute into our recurrence
    $$a_n = 28a_{n-2} - 3a_{n-1}$$
    $$c \cdot r^n = 28 \cdot c \cdot r^{n-2} - 3 \cdot c \cdot r^{n-1}$$

    This implies
    $$c \cdot r^n + 3 \cdot c \cdot r^{n-1} - 28 \cdot c \cdot r^{n-2} = 0$$
    $$r^2 + 3r - 28 = 0 \quad \text{(dividing by } c \cdot r^{n-2}\text{)}$$
    $$(r-4)(r+7) = 0$$

    This tells us that $r = 4$ or $r = -7$. This gives us a general solution of
    $$a_n = c_1 4^n + c_2 (-7)^n.$$

    Now we use our initial conditions to find c_1 and c_2. We note that
    $$2 = a_0 \quad = c_1 4^0 + c_2 (-7)^0 = c_1 + c_2$$
    $$19 = a_1 \quad = c_1 4^1 + c_2 (-7)^1 = 4c_1 - 7c_2$$

    This gives us a system of linear equations which we can solve to obtain $c_1 = 3$ and $c_2 = -1$. Thus our closed form for a_n is
    $$a_n = 3 \cdot 4^n - (-7)^n. \qquad \square$$

27. Define a graph Q_k for $k \geq 1$ to be the "k-cube graph". Each vertex of Q_k corresponds to some length k binary string. Two vertices are adjacent if and only if their strings differ in exactly 1 coordinate. How many vertices does the graph Q_k have? How many edges does Q_k have? (Note: your answers to both of these questions should be functions of k.)

    Solution. Each vertex of Q_k represents a binary string of length k. There are 2^k binary strings of length k, and thus 2^k vertices of Q_k.

    The degree of any vertex in our graph will be exactly k (there is one edge for each coordinate being flipped). By the Handshaking Lemma, we know
    $$\sum_{v \in V(Q_k)} d(v) = \sum_{v \in V(Q_k)} k = k \cdot 2^k = 2|E(Q_k)|$$
    so the number of edges in Q_k is $k \cdot 2^{k-1}$. $\qquad \square$

28. Show that the number of vertices of odd degree in a graph must be even.

   Solution. Recall that the Handshaking Lemma states that
   $$\sum_{v \in V} d(v) = 2|E|.$$
   In particular, this means that if we sum up the degree of all vertices in the graph, we get an even number. Since the sum of all the even degrees will be even, this implies the sum of the odd degrees must also be even in order for the total sum to be even. But this means we must sum up an even number of odd degrees, and since each degree corresponds to a vertex in our graph, we conclude that the number of vertices of odd degree in a graph must be even. □

29. Give a formula for $\chi(\overline{K_n}; k)$ and $\chi(K_n; k)$. Note that $\overline{K_n}$ indicates the graph on n vertices with no edges. You may assume $k \geq \chi(K_n)$.

   Solution. Because there are no edges in $\overline{K_n}$, there are no restrictions on the colors of the vertices; that is, each vertex may receive any of the k possible colors. Thus, $\chi(\overline{K_n}; k) = k^n$.

   Now consider the complete graph K_n. Let the vertices of our K_n be $v_1, v_2, \ldots, v_n$. We will assign colors to the vertices in order. We can assign any of the k colors to v_1. Since v_2 is adjacent to v_1, it cannot be the same color, so we have $k-1$ options for coloring v_2. Similarly, v_3 is adjacent to both v_2 and v_1 so there are $k-2$ allowed colors for its assignment. This pattern continues, yielding
   $$\chi(K_n; k) = k \cdot (k-1) \cdot (k-2) \cdots (k-n+1) = \frac{k!}{(n-k)!}. \qquad \square$$

30. A graph $G = (V, E)$ is called *bipartite* if there exists some partition X, Y of V (i.e., $X \cup Y = V$ and $X \cap Y = \emptyset$) such that every edge of G has one endpoint in X and one endpoint in Y. Explain why every bipartite graph is 2-colorable.

   Solution. We claim that coloring all vertices of X with one color (say red) and all vertices of Y with a second color (say blue) gives a proper 2-coloring of a bipartite graph. Since any edge in a bipartite graph must have exactly one endpoint in X and one endpoint in Y, every edge of our graph has one red and one blue endpoint. Thus, no two adjacent vertices are the same color, so this is a proper 2-coloring, implying any bipartite graph is 2-colorable. □

31. Let b_n the number of ways to write a positive integer n as a sum of nonnegative powers of 2, where by convention we set $b_0 = 1$. Find the generating function for this sequence and use it to prove that $b_n = \sum_{k=0}^{[\frac{n}{2}]} b_k$.

Solution. This is a standard example for using products of generating functions. The sums of copies of 2^j correspond to the generating function $1 + X^{2^j} + X^{2 \cdot 2^j} + X^{3 \cdot 2^j} + \ldots$. Hence the generating function for the sequence (b_n) is

$$F(X) = \sum_{k=0}^{\infty} b_k X^k = \prod_{j=0}^{\infty} (1 + X^{2^j} + X^{2 \cdot 2^j} + X^{3 \cdot 2^j} + \ldots).$$

Now obviously $1 + X^{2^j} + X^{2 \cdot 2^j} + X^{3 \cdot 2^j} + \ldots = \dfrac{1}{1 - X^{2^j}}$, so

$$F(X) = \prod_{j=0}^{\infty} \frac{1}{1 - X^{2^j}}.$$

Now let us denote by $a_n = \sum_{k=0}^{[\frac{n}{2}]} b_k$. The trick is to prove that the generating function of a_n is the same as that of b_n.
Thus consider

$$G(X) = \sum_{n \geq 0} a_n X^n = \sum_{n \geq 0} \left(\sum_{k=0}^{[\frac{n}{2}]} b_k \right) X^n.$$

Switching the order of sums, we have

$$G(X) = \sum_{k \geq 0} b_k (X^{2k} + X^{2k+1} + \ldots) = \sum_{k \geq 0} b_k \frac{X^{2k}}{1 - X}.$$

Thus we see that

$$G(X) = \frac{1}{1 - X} \sum_{k \geq 0} b_k X^{2k} = \frac{F(X^2)}{1 - X} = F(X). \qquad \square$$

32. Let n be a positive integer. Show that the number of partitions of n, where each part appears at least twice, is equal to the number of partitions of n into parts that are divisible by 2 or 3.

Solution. Again we are interested the two sequences considered have the same generating function. Let a_n be the number of partitions of n where each part appears at least twice and b_n the number of partitions of n where each part is divisible by 2 or 3.

Also let $F(X) = \sum_{n \geq 0} a_n X^n$ and $G(X) = \sum_{n \geq 0} b_n X^n$.

According to the hypothesis

$$F(X) = \prod_{k=1}^{\infty}(1 + X^{2k} + X^{3k} + \ldots + X^{nk} + \ldots)$$

and thus we can rewrite as

$$F(X) = \prod_{k=1}^{\infty}(1 + X^k + X^{2k} + X^{3k} + \ldots - X^k) = \prod_{k=1}^{\infty}\left(\frac{1}{1-X^k} - X^k\right)$$

We obtain the equality

$$F(X) = \prod_{k=1}^{\infty} \frac{X^{2k} - X^k + 1}{1 - X^k}.$$

For the second generating function we can at first write as

$$G(X) = \prod_{k=1}^{\infty} \frac{\left(\sum_{i=0}^{\infty} X^{2ki}\right)\left(\sum_{i=0}^{\infty} X^{3ki}\right)}{\sum_{i=0}^{\infty} X^{6ki}}$$

since when we take multiples of 2 or 3 on top we are overcounting the multiples of 6.

Now again we can rewrite

$$G(X) = \prod_{k=1}^{\infty} \frac{1 - X^{6k}}{(1-X^{2k})(1-X^{3k})}$$

$$= \prod_{k=1}^{\infty} \frac{(1-X^{3k})(1+X^{3k})}{(1-X^{2k})(1-X^{3k})}$$

$$= \prod_{k=1}^{\infty} \frac{1+X^{3k}}{1-X^{2k}}.$$

The fact that $F(X) = G(X)$ now boils down to the identity

$$1 + a^3 = (1+a)(1-a+a^2)$$

applied to each $a = X^k$ in the infinite product. □

Solutions for Introductory Problems 179

33. Let $f(n,k)$ be the number of ways of distributing k candies to n children so that each child receives at most 2 candies. For example $f(3,7) = 0$, $f(3,6) = 1$, $f(3,4) = 6$. Determine the value of

$$f(2006,1) + f(2006,4) + \ldots + f(2006,1000)$$
$$+ f(2006,1003) + \ldots + f(2006,4012).$$

(Adapted Canada 2006)

Solution. Note that $f(n,k)$ is the number of solutions $(x_1, x_2, \cdots, x_n)$ to $x_1 + x_2 + \cdots + x_n = k$ where $x_i \in \{0,1,2\}$ for $1 \le i \le n$. Equivalently, $f(n,k)$ is the coefficient of X^k in $G(X) = (1 + X + X^2)^n$.

Let ω be a primitive third root of unity. Note that

$$\frac{1}{3}[G(X) + \omega^2 G(\omega X) + \omega G(\omega^2 X)] = f(n,1)X + f(n,4)X^4 + f(n,7)X^7 + \cdots$$

Plugging in $X = 1$ we get

$$\frac{1}{3}[G(1) + \omega^2 G(\omega) + \omega G(\omega^2)] = f(n,1) + f(n,4) + f(n,7) + \cdots$$

Now it is easy to see that $G(1) = 3^n, G(\omega) = 0, G(\omega^2) = 0$ so the desired sum is simply 3^{n-1}. The answer to the problem is thus 3^{2005}.

Alternate Solution. Note that, after unwinding definitions, the requested sum is the number of ways to distribute a number of candies which is 1 mod 3 to the 2006 children. For this we use the Product Rule, going through the children in order. For the first child, there are 3 possibilities, we give him 0, 1, or 2 candies. □

34. How many n digit numbers are there such that they are divisible by 3 and all their digits are either $2, 3, 7, 9$?

(Romania)

Solution. Let us consider the following generating function

$$F(X) = \sum_{a+b+c+d=n} X^{2a+3b+7c+9d}$$

which quantifies the numbers we have since our number has a digits equal to 2, b digits equal to 3, c digits equal to 7 and d digits equal to 9. Note that actually

$$F(X) = (X^2 + X^3 + X^7 + X^9)^n.$$

Also from the initial expression we need something that detects divisibility by 3 in the exponent, and again we are led to consider a primitive root of unity of order 3, call it ω.

Then $F(\omega) = A + B\omega + C\omega^2$ where A represents the number of n digit numbers divisible by 3, B is those which are congruent to 1 modulo 3 and finally C is coming from numbers which are congruent to 2 modulo 3.

From our observation

$$F(\omega) = (\omega^2 + \omega^3 + \omega^7 + \omega^9)^n = (2 + \omega + \omega^2)^n = 1,$$

since $1 + \omega + \omega^2 = 0$.

Thus $A - 1 + B\omega + C\omega^2 = 0$ and so ω is a root of the polynomial $(A-1) + BT + CT^2$, and since the minimal polynomial of ω is $T^2 + T + 1$ it follows that

$$T^2 + T + 1 \mid (A-1) + BT + CT^2$$

and since they have the same degree we must have $A - 1 = B = C$. To finish up note that $A + B + C = F(1) = 4^n$ so $A = \dfrac{4^n + 2}{3}$. □

35. Consider n points $P_1, P_2, \ldots, P_n$ lying on a straight line. We color each point in white, red, green, blue and violet. A coloring is admissible if for each pair of consecutive points P_i, P_{i+1} ($i = 1, 2, \ldots, n-1$) either both points have the same color, or at least one of them is white. How many admissible collorings are there?

(Austrian-Polish 1998)

Solution. Let w_n the number of those collorings for which P_n is white, r_n the number of those collorings where it is red, g_n where it is green, b_n where it is blue and finally v_n where it is violet. We are interested in a recurrence for $s_n = w_n + r_n + g_n + b_n + v_n$.

Obviously if P_n is white then P_{n-1} can have any color so $w_n = s_{n-1}$.

If P_n is not white then we have the following recurrences

$$r_n = w_{n-1} + r_{n-1}$$

$$g_n = w_{n-1} + g_{n-1}$$

$$b_n = w_{n-1} + b_{n-1}$$

$$v_n = w_{n-1} + v_{n-1}$$

Thus summing we obtain $s_n = s_{n-1} + 3w_{n-1} + s_{n-1} = 2s_{n-1} + 3s_{n-2}$. The characteristic equation is $r^2 - 2r - 3 = 0$ with roots $r_1 = 3$ and $r_2 = -1$.

It follows that $s_n = C_1 3^n + C_2(-1)^n$. Since $s_1 = 5$ and $s_2 = 13$, solving we have $C_1 = \dfrac{3}{2}$ and $C_2 = -\dfrac{1}{2}$ thus

$$s_n = \frac{3^{n+1} + (-1)^{n+1}}{2}.$$
$\square$

36. At a summer camp there are n girls, $G_1, G_2, \ldots, G_n$ and $2n-1$ boys $B_1, B_2, \ldots, B_{2n-1}$. Girl G_i knows boys $B_1, B_2, \ldots, B_{2i-1}$ and no others. Prove that the number of ways to choose r boy-girl pairs so that each girl in the pair knows the boy in the pair is $\dbinom{n}{r} \dfrac{n!}{(n-r)!}$.

(Czech-Slovak Match 1998)

Solution. Let $P(n,r)$ the number of such pairings. We seek to obtain a recurrence relation.

If the girl G_n is not in the pairs then we simply have $P(n-1, r)$.

If the girl G_n in included in the pairs, then obviously the other $r-1$ girls can be paired with boys in $P(n-1, r-1)$ ways and she will still have $2n-r$ boys to choose from to make a a pair. Thus we have $(2n-r)P(n-1, r-1)$ such pairings.

Thus $P(n,r) = P(n-1, r) + (2n-r)P(n-1, r-1)$.

To finish up it suffices to show that we have agreement for $r = 1$ and that the two sequences satisfy the same recurrence.

For $r = 1$ we have $P(n, 1) = 1 + 3 + 4 + \ldots + 2n - 1 = n^2$ and they agree and we leave it up to the reader to check the identity involving the binomial coefficients.
$\square$

37. Let p be a positive integer, $p > 1$. Find the number of $m \times n$ tables with entries in the set $\{1, 2, \ldots, p\}$ and such that the sum of elements in each row and each column is not divisible by p.

(Mediterranean 2010)

Solution. Let's solve the $1 \times n$ case first. We need to find the number of sequences of n terms from $1, 2, \ldots p-1$ whose sum is not divisible by p. Let a_n be the number of such sequences.

We claim the recurrence $a_n = (p-2)a_{n-1} + (p-1)a_{n-2}$ holds. Suppose we have an n-term sequence. If the sum of the first $n-1$ terms of this sequence is not a multiple of p, say it is $s \pmod{p}$, then we can pick the first $n-1$ terms in a_{n-1} ways and the last term can be any residue mod p except 0 and $-s$, for $(p-2)a_{n-1}$ sequences. If the sum of the first $n-1$ terms is divisible by p, then the sum of the first $n-2$ terms is not a multiple of p. Hence there are a_{n-2} ways to pick these terms. The $(n-1)$-st term must be the unique value that gives a multiple of p for the sum of the first $n-1$ terms, and the last term can be any of the $p-1$ nonzero values. Thus this case gives $(p-1)a_{n-2}$ sequences. Summing these two cases up gives the desired recurrence.

It is trivial to see that $a_1 = p-1$ and $a_2 = (p-1)(p-2)$. Combining this with the recurrence, which has characteristic polynomial

$$r^2 - (p-2)r - (p-1) = 0,$$

and thus roots $r_1 = -1$ and $r_2 = p-1$ we obtain

$$a_n = \frac{p-1}{p}\left((p-1)^n - (-1)^n\right).$$

The reader may notice the similarity between this answer and the answers to Examples 74 and 91. This part of the current problem, and those two exercises are all essentially the same problem in different guises.

Now let's proceed to the general case. We can consider the top right $(m-1) \times (n-1)$ subgrid of the table and fill it arbitrarily with numbers. There are $p^{(m-1)(n-1)}$ possible ways to do this.

For any of these choices, let the bottom row have entries $x_1, x_2, \ldots, x_{n-1}, z$, and the last column have entries $y_1, \ldots, y_{m-1}, z$, where z is the common element in the corner. We claim that we can construct a bijection to $1 \times (m+n-1)$ good tables. If we prove this, we will be done.

To do this let's denote by $r_1, r, \ldots, r_{n-1}$ the sums of the rows of the $(m-1) \times (n-1)$ subgrid and by $c_1, \ldots, c_{m-1}$ the sums of the columns of this subgrid. Note that $s = r_1 + \ldots + r_{n-1} = c_1 + \ldots + c_{m-1}$. Let us also denote by $z^* = y_1 + \ldots + y_{m-1} + z$. Our map will send $(x_1, \ldots, x_{n-1}, y_1, \ldots, y_{m-1}, z)$ to the residues modulo p of

$$(x_1 + r_1, \ldots, x_{n-1} + r_{n-1}, -(y_1 + c_1), \ldots, -(y_{m-1} + c_{m-1}), z^*).$$

The checking is easy. Each component is nonzero because each is either a row sum of the full table or the negative of a column sum, hence nonzero

Solutions for Introductory Problems   183

modulo p. The last step is to note that the sum of all the components is $x_1 + \ldots + x_{n-1} + z$, which is also nonzero since it is the final column sum. Also note that $r_1, r, \ldots, r_{n-1}$ and $c_1, \ldots, c_{m-1}$ are fixed, so we can invert the map in an obvious way.

Thus we obtain in total

$$p^{(m-1)(n-1)-1}(p-1)\left((p-1)^{m+n-1} - (-1)^{m+n-1}\right) \text{ tables.} \qquad \square$$

38. Let $\mathcal{F}$ be a family of subsets of the set $\{1, 2, \ldots, n\}$ such that every element in $\mathcal{F}$ has cardinality 3 and moreover for any two distinct elements $A, B \in \mathcal{F}$ we have $|A \cap B| \leq 1$. Prove that

$$|\mathcal{F}| \leq \frac{n(n-1)}{6}.$$

Solution. Consider the collection $\mathcal{C}$ of 2-element subsets of any subset from $\mathcal{F}$.

From the hypothesis we know that for any two distinct A and B from $\mathcal{F}$ we have $|A \cap B| \leq 1$, so the 2-element subsets of these have to be distinct.

Thus $|\mathcal{C}| = 3|\mathcal{F}|$, since every 3-element set has three 2-element subsets.

On the other hand $\mathcal{C}$ is contained in the family of all 2-element subsets of $\{1, 2, \ldots, n\}$, thus $|\mathcal{C}| \leq \binom{n}{2}$.

Thus we obtain $3|\mathcal{F}| \leq \binom{n}{2}$ and so $|\mathcal{F}| \leq \frac{n(n-1)}{6}$. $\qquad \square$

39. Let S be a set of n persons such that:

   (a) Any person is acquainted to exactly k other persons in S.

   (b) Any two persons that are acquainted have exactly l common acquaintances in S.

   (c) Any two persons that are not acquainted have exactly m common acquaintances in S.

   Prove that $m(n-k) - k(k-l) + k - m = 0$.

Solution. Rearrange the required equality as $m(n-k-1) = k(k-l-1)$.

For a given person p consider the tuples of the type (p, A, B) such that p knows A and p knows B but A does not know B.

At first glance we can choose A in k ways, and since p knows B but A does not know B, it follows that B is not a common acquaintance of p and A and thus we can choose B in $k-l-1$ ways from the acquaintances of p.

Thus we have a total of $k(k-l-1)$ tuples. Now sum it over over all p in S to obtain $nk(k-l-1)$ such triples.

Now let's count in a different way, starting from B. We can choose B from S in n ways, then since A does not know B, we can choose A out of $n-k-1$ person, and since p is a common acquaintance of A and B this can be done out of m possibilities. Thus we get a total of $n(n-k-1)m$ of triples. Putting together the two different counts we must have $nk(k-l-1) = n(n-k-1)m$ and the required equality follows. $\square$

40. A school has n students, and each student can take any number of classes. Every class has at least two students in it. We know that if two different classes have at least two common students, then the number of students in these two classes is different. Prove that the number of classes is not greater that $(n-1)^2$.

(Iran 2010)

Solution. For each $2 \leq k \leq n$ let a_k be the number of classes that have precisely k students. Then the total number of classes is $\sum_{k=2}^{n} a_k$.

Now, let us look at pairs of students who share a class with precisely k students. By hypothesis, no pair of students share two classes of size k, so there are $a_k \binom{k}{2}$ such pairs. However, there are at most $\binom{n}{2}$ such pairs, since there are only this many pairs of students. Thus we have the inequality $a_k k(k-1) \leq n(n-1)$ or $a_k \leq \dfrac{n(n-1)}{k(k-1)}$.

Summing up we obtain that the number of classes is at most

$$n(n-1) \sum_{k=2}^{n} \frac{1}{k(k-1)} = n(n-1) \sum_{k=2}^{n} \left(\frac{1}{k-1} - \frac{1}{k} \right)$$

$$= n(n-1)\left(1 - \frac{1}{n}\right) = (n-1)^2. \quad \square$$

41. There are 10001 students at an university. Some students join together to form several clubs (a student may belong to different clubs). Some clubs

Solutions for Introductory Problems

join together to form several societies (a club may belong to different societies). There are a total of k societies. Suppose that the following conditions hold:

(a) Each pair of students are in exactly one club.

(b) For each student and each society, the student is in exactly one club of the society.

(c) Each club has an odd number of students. In addition, a club with $2m+1$ students (m is a positive integer) is in exactly m societies.

Find all possible values of k.

(IMO Shortlist 2004)

Solution. Let $\mathcal{C}_1$,..., $\mathcal{C}_n$ be the clubs at the university and let $2c_i+1$ be the number of students in the club $\mathcal{C}_i$, for $1 \leq i \leq n$.

Each club has $\binom{2c_i+1}{2}$ pairs of students. Thus $\sum_{i}^{n}\binom{2c_i+1}{2}$ gives us the total number of pairs of students. Thus we have the equality

$$\sum_{i=1}^{n}\binom{2c_i+1}{2} = \binom{10001}{2}.$$

Now let us count something two ways. Condition (b) tells us that each society has precisely 10001 members. Thus the number of pairs of a student and a society he belongs to is equal to $10001k$, where k is the number of societies.

We can also count these pairs using the clubs in each society. The last condition, (c), tells us that it is also equal to $\sum_{i=1}^{n}(2c_i+1)c_i$.

Thus we have

$$\sum_{i=1}^{n}(2c_i+1)c_i = 10001k.$$

Putting together the two identities we've obtained, we have

$$10001k = 10001 \cdot 5000 \quad \text{so} \quad k = 5000.$$

Such a distribution can be attained by taking one club to have 10001 students and all 5000 societies to consist of just this club. We may also add any number of clubs with only 1 student belonging to no society. □

42. Each member of a club has at most three enemies in the club. (Here enemies are mutual.) Show that the members can be divided into two groups so that each member in each group has at most one enemy in the group.

Solution. Fix a division of the club members into two groups. Call a pair of people bad if they are in the same group and are enemies. Since there are only finitely many divisions of the club, we can choose one that minimizes the number of bad pairs. We claim that in this division no one will have two enemies in his group.

To see this suppose Bob has two enemies in his group. Then Bob has at most one enemy in the other group. Hence if we move Bob to the other group, we will lower the number of bad pairs. This contradicts our minimallity assumption, hence no such member exists. $\square$

43. Several positive integers are written on a blackboard. One can erase any two distinct integers and write their greatest common divisor and least common multiple instead. Prove that eventually the numbers will stop changing.

(St Petersburg, 1996)

Solution. The first observation is that the product of the numbers written on the board does not change. This is a consequence of the identity $(a,b) \cdot [a,b] = ab$, where we've used the standard notations for the greatest common divisor and lowest common multiple.

This is not enough to solve the problem. Let us look at $S = \sum_{i=1}^{n} a_i$ where a_i are the numbers written on the board.

We claim that S increases with each operation; namely we have to prove that $(a,b) + [a,b] \geq a + b$.

For this let $a = pq$, $b = pr$, where $p = (a,b)$ and thus $[a,b] = pqr$. The inequality becomes $p + pqr \geq pq + pr$ and thus we should have $1 + rq \geq r + q$ which is equivalent to $(r-1)(q-1) \geq 0$ and this is obvious.

On the other hand we can use a crude bound $S \leq n \prod_{i=1}^{n} a_i$ and this quantity on the right hand side is invariant under the operations done on the board.

We can conclude thus that since S increases and is bounded, it must at one point stop changing and thus the numbers on the board stop changing. □

44. Which single squares can be removed from a 7×7 board so that the rest can be tiled with 1×3 trominos?

    Solution. We produce the following two labellings of the cells in 7×7 board

0	1	2	0	1	2	0
1	2	0	1	2	0	1
2	0	1	2	0	1	2
0	1	2	0	1	2	0
1	2	0	1	2	0	1
2	0	1	2	0	1	2
0	1	2	0	1	2	0

0	1	2	0	1	2	0
2	0	1	2	0	1	2
1	2	0	1	2	0	1
0	1	2	0	1	2	0
2	0	1	2	0	1	2
1	2	0	1	2	0	1
0	1	2	0	1	2	0

    where think about $0, 1, 2$ as modulo 3. No matter how we place a tromino on the board, in either labeling we have that the sum of the labels inside it is 0 modulo 3.

    Since the sum of the the labels in the cells, in both labellings, is equal to 0 modulo 3; if such a covering exists we must have removed a cell labeled with 0 in both cases.

    Thus the removed cell must be in one of the following cell positions (starting from top left): $(1,1)$, $(1,4)$, $(1,7)$, $(4,1)$, $(4,4)$, $(4,7)$, $(7,1)$, $(7,4)$, $(7,7)$.

    Finally, note that if any of these cells is removed, then it is easy to find a tiling with 1×3 trominos. □

45. An $m \times n$ array of real numbers is given. When the sum of the numbers in any row or column is negative, we may switch the signs of all the numbers in that row or column. If this operation is iterated, prove that all of the row or column sums eventually become nonnegative.

    (Russia)

    Solution. Let S be the sum of all the mn numbers in the table. Note that after an operation, each number stays the same or turns to its negative. Hence there are at most 2^{mn} possible tables. So S can only have finitely many possible values.

Look for a line (a row or a column) whose numbers have a negative sum. If no such line exists, then we are done. Otherwise, reverse the signs of all the numbers in that line. Then S will increase. Since S can take on only finitely many values, S can increase only finitely many times. So eventually the sum of the numbers in every line must be nonnegative. $\square$

46. For a $n \times n$ table with positive integer entries we are allowed to make the following operations: multiply each entry of a row by 2 or subtract 1 from each number in a column. Prove that we can always reach a table where are all the entries are 0.

Solution. We will show that it is possible to make any column into a column of all ones, keeping all entries positive. If one has done this, then subtracting one from that column will leave a column of all zeroes. Once it has all zeroes no row operation will change it and neither will a column operation on a different column, so we can effectively reduce to the case of a grid with one fewer column.

We will show that by row operations and column operations on the last column, we can make it into the all ones column. Over all combinations of these operations that give only positive numbers in this column, consider the outcome for which the sum of the elements is lowest. Let $c_1, c_2, \ldots, c_n$ be the resulting entries.

First, note that we must have some 1's among these entries. Otherwise, we could subtract a 1 from each entry of the last column and thus get a lower sum. Assume without loss of generality, that $c_1, \ldots, c_k$ are equal to 1, with $k < n$. Now, multiply the first k rows by 2 and subtract one from the last column.
Thus our entries would have become $(1, \ldots, 1, c_{k+1} - 1, \ldots, c_n - 1)$ which has a lower sum, a contradiction.

Thus we must end with the last column consisting only of 1's and subtracting a 1 from each entry would make it all zeroes. Hence we can drop this column. Since we have not performed a column operation on any other column, all the remaining entries will still be positive. $\square$

47. Alfred and Bonnie play a game in which they take turns tossing a fair coin. The winner is the first person to obtain a head. They play this game several times, with the stipulation that the loser of a game goes first in the next game. Suppose Alfred goes first in the first game, what is the probability that he wins the 6th game?

(AIME 1993)

Solution. Since play always alternates between Alfred and Bonnie, we can imagine that they alternate flipping a coin, and we ask for the probability that Alfred gets the 6th heads overall.

In general, let a_n be the probability that the first player gets the n-th heads overall. We will write down a recursion for a_n. Suppose the first player, gets a tail (which occurs with probability $1/2$). Then he effectively becomes the second player. Hence the probability that he gets the n-th head is $1 - a_n$. Suppose the first player gets a head (also probability $1/2$). If $n = 1$, then he gets the first head and wins. If $n > 1$, then he effectively becomes the second player and the contest becomes to get the $(n-1)$-st head. Hence he wins with probability $1 - a_{n-1}$. Note that we can combine these two cases if we define $a_0 = 0$. Thus we get

$$a_n = \frac{1}{2}(1 - a_n) + \frac{1}{2}(1 - a_{n-1}),$$

or after rearranging

$$a_n = \frac{2 - a_{n-1}}{3}.$$

To solve this we let $x_n = a_n - \frac{1}{2}$. Then $x_0 = -\frac{1}{2}$ and the recursion becomes $x_n = -\frac{x_{n-1}}{3}$. Hence

$$x_n = \frac{(-1)^{n-1}}{2 \cdot 3^n} \quad \text{and} \quad a_n = \frac{1}{2} + \frac{(-1)^{n-1}}{2 \cdot 3^n}.$$

In particular,

$$a_6 = \frac{364}{729}. \qquad \square$$

48. Let $A_1, A_2, \ldots, A_k$ be subsets of $\{1, 2, \ldots, n\}$, each with three elements. Show that it is possible to color the elements of $\{1, 2, \ldots, n\}$ with c colors, such that at most $\dfrac{k}{c^2}$ of the A_i's are monochromatic.

Solution. Look at random colorings, where we assign each of the c colors to an element with equal probability $1/c$ and with different elements colored independently. Let X be the random variable whose value is the number of monochromatic sets among the A_i.

The probability that a set is monochromatic is obviously equal to $\dfrac{1}{c^2}$, since there are only c ways to monochromatically color a set with 3 elements, and there are c^3 ways to color it.

Thus $E[X] = \dfrac{k}{c^2}$ since there are k sets and thus there must be a coloring with at most $\dfrac{k}{c^2}$ of the A_i's monochromatic. □

49. Consider a set S with n elements. Let $A_1, A_2, \ldots, A_{n+1}$ be non-empty distinct subsets of S. Then
$$\sum_{1 \le i < j \le n} \frac{|A_i \cap A_j|}{|A_i| \cdot |A_j|} \ge 1.$$

Solution. Suppose we choose a random element X_i of A_i, where each element of A_i is equally likely to be chosen and our choices for different indices are independent. Then there are $|A_i|$ possible choices for X_i and $|A_j|$ possible choices for X_j and there are $|A_i \cap A_j|$ ways to choose them to be equal. Hence
$$P(X_i = X_j) = \frac{|A_i \cap A_j|}{|A_i| \cdot |A_j|}.$$

Let Y be the random variable whose value is the number of pairs (i, j) with $i < j$ and $X_i = X_j$. Then by linearity of expectation we have
$$E[Y] = \sum_{1 \le i < j \le n} \frac{|A_i \cap A_j|}{|A_i| \cdot |A_j|}.$$

Since overall we chose $n+1$ elements of S, the Pigeonhole Principle implies two of them must be equal. Hence $Y \ge 1$ for any choice and $E[Y] \ge 1$, which is exactly the desired inequality. □

50. Let a_j, b_j, c_j be integers for $1 \le j \le N$. Assume for each j, at least one of a_j, b_j, c_j is odd. Show that there exists integers r, s, t such that $ra_j + sb_j + tc_j$ is odd for at least $\dfrac{4N}{7}$ of the values of j, $1 \le j \le N$.

(Putnam 2000)

Solution. We begin with an easy observation. We will work modulo 2. Suppose that $a, b, c \in \mathbb{Z}/2\mathbb{Z}$ are not all zero. Then there are four choices of $r, s, t \in \mathbb{Z}/2\mathbb{Z}$ such that $ra + sb + tc = 1$.

Now choose $r, s, t \in \mathbb{Z}/2\mathbb{Z}$ randomly (uniformly) not all zero, and let $X_1, X_2, \ldots, X_N$ be random variables for the events $ra_i + sb_i + tc_i = 1$, where X_i is 1 if $ra_i + sb_i + tc_i = 1$, and 0 otherwise.

Then since there are 7 choices for (r, s, t) and using the observation at the beginning we have that $E[X_i] = \frac{4}{7}$. By using linearity of expectation

$$E[X_1 + \ldots + X_N] = E[X_1] + \ldots + E[X_N] = \frac{4N}{7}.$$

As $X_1 + \ldots + X_N$ is a random variable giving the number of possible values for i such that $ra_i + sb_i + tc_i = 1$, we immediately conclude that there is a choice of r, s, t that gives at least $\frac{4N}{7}$ values of i such that $ra_i + sb_i + tc_i = 1$ in $\mathbb{Z}/2\mathbb{Z}$, as desired. □

51. What is the probability that n random points on a circle are contained in a semicircle?

    Solution. Suppose the points are $x_1, \ldots, x_n$. Call a point x_i a leftmost point if $x_j \in [x_i, x_i + \pi)$ for all j (addition is modulo 2π). Note that the collection of n random points will be contained in a semicircle if and only if there is a leftmost point. For any pair i, j of distinct indices, the probability that $x_j \in [x_i, x_i + \pi)$ is obviously $1/2$. Thus the probability that x_i is a leftmost point is $\frac{1}{2^{n-1}}$.

    Note that if $x_j \in [x_i, x_i + \pi)$, then $x_i \notin [x_j, x_j + \pi)$. Thus for any configuration there is at most one leftmost point. Thus the expected number of leftmost points is the same as the probability that there is a leftmost point. Since there are n points each with a probability of $\frac{1}{2^{n-1}}$ of being leftmost, we get that the probability is $\frac{n}{2^{n-1}}$. □

52. A broken line with length greater than 1000 lies inside a unit square. Prove that there exists a line which intersects the broken line in at least 501 points.

    Solution. Denote by $A_1, \ldots, A_n$ the vertices of the broken line, let $P_1, \ldots, P_n$ be the projections on a horizontal side of the square and let $R_1, \ldots, R_n$ be the projections on a vertical side.

    By the triangle inequality,

    $$A_i A_{i+1} \leq P_i P_{i+1} + R_i R_{i+1}.$$

Thus
$$\sum_{i=1}^{n-1} P_i P_{i+1} + \sum_{i=1}^{n-1} R_i R_{i+1} > 1000.$$

Thus we can assume without loss of generality that
$$\sum_{i=1}^{n-1} P_i P_{i+1} > 500.$$

Since the union of the segments $P_i P_{i+1}$ has total length greater than 500 and all these segments lie on a side of the unit square, there must some point that is in at least 501 of them. Call it B. Then the vertical line through B intersects at least 501 of the segments $A_i A_{i+1}$. □

53. Prove that 2015 points, no three collinear, in the plane determine at least 403 convex quadrilaterals, which have disjoint interiors.

    Solution. There are only finitely many lines passing through two of the 2015 points, so we can choose a line ℓ which is not parallel to any of them. Regard the line ℓ as being horizontal. Then the 2015 points all have different y-coordinates. Hence we can assume they are ordered $P_1, \ldots, P_{2015}$ in increasing order of y-coordinates. Break these points into 403 sets of 5 so that the k-th set is $\{P_{5k-4}, P_{5k-3}, P_{5k-2}, P_{5k-1}, P_{5k}\}$. By Example 110, we can pick 4 vertices from each of these sets which are vertices of a convex quadrilateral.

    It remains to show that these quadrilaterals are disjoint. But this is easy. If $j < k$, then since the y-coordinates are increasing, we can choose a horizontal line m which is above P_{5j}, the highest point in the first set, and below P_{5k-4}, the lowest point in the second. Hence the j-th convex quadrilateral is below m and the k-th is above m. Thus they cannot meet. □

54. Given $2n$ distinct points, no three collinear, in the plane, if we color n of them with red and n of them with blue, prove that we can connect each blue with a red point such that the pairwise segments are nonintersecting.

    Solution. Consider all possible pairings of red and blue points. For each pairing $\mathcal{P}$, we define the sum $S(\mathcal{P}) = \sum_{i=1}^{n} R_i B_i$, where R_i and B_i are the red points and blue points, respectively, and paired points have the same index.

Since the number of pairings is finite, we can consider a pairing such that the sum for that pairing is minimal. Let's show that this pairing satisfies the conditions of the problem.

Assume the contrary. This would mean that there exist four points $A = B_i$, $B = B_j$, $C = R_i$, and $D = R_j$ such that AC intersects BD. Thus the quadrilateral $ABCD$ is convex.

It follows easily from the triangle inequality, that in this case $AD+BC < AC+BD$. Define a new pairing by making $C = R_j$ and $D = R_i$, keeping the other points the same. This would lead to a new pairing with a lower sum, and thus we have a contradiction. $\square$

55. Suppose that a set S in the plane containing n points has the property that any three points can be covered by an infinite strip of width 1. Prove that S can be covered by a strip of width 2

(Balkan Math Olympiad 2010)

Solution. Let's start like in Example 107. Consider the triangle ABC with vertices in S of maximal area. Let D, E, F be the points such A is the middpoint of EF, B is the midpoint of FD, and C is the midpoint of DE. We know from the argument in Example 107, that the set S is contained completely inside the triangle DEF.

By the hypothesis, $\triangle ABC$ can be covered by a strip of width 1. Let G be the centroid of $\triangle ABC$, and consider the homothety centered at G with ratio -2. This will send triangle $\triangle ABC$ to triangle $\triangle DEF$ and the strip of width 1 containing $\triangle ABC$ will be sent to a strip of width 2 containing $\triangle DEF$. Thus we are done since S is contained in triangle $\triangle DEF$.

Alternate Solution.

Lemma. The minimum width w of a strip containing triangle $\triangle ABC$ is the length h of the altitude to the longest side.

Proof. It is easy to see that triangle $\triangle ABC$ is contained in a strip of width h. Merely, take the strip between the line through the longest side and the parallel to that side through the opposite vertex. Hence $w \leq h$.

Suppose $\triangle ABC$ is contained in a strip of width w between parallel lines ℓ and m. Without loss of generality, we may assume the projection of B onto ℓ lies between the projections of A and C onto ℓ. Then there is a vertical segment through B of length at most w that meets AC. Since the altidude from B to AC is the shortest segment joining B to AC, we have $w \geq h_B$. Since the altitude to the longest side is the shortest of the three altitudes, we have $w \geq h$. Combining these we see that $w = h$.

Solving the problem with this lemma is easy. Let A and B be points in S at a maximum distance apart. Then for any point C in S, the side AB is the longest side of triangle $\triangle ABC$, and hence by the lemma, C lies at distance at most 1 from the line AB. Thus S lies in the strip of width two whose midline is the line AB. □

56. Let S be a set of points, no three collinear, with at least three points, such that for any distinct $A, B, C \in S$ the circumcenter of $\triangle ABC$ is in S. Prove that S is infinite.

Solution. Assume the contrary. Consider the triangle $\triangle ABC$ with vertices in S of minimum circumradius. Let R be the circumradius of $\triangle ABC$ and assume without loss of generality that $\angle A \geq \angle B \geq \angle C$.

We have two cases

a) The triangle $\triangle ABC$ is obtuse or right. Let O the circumcenter. Since $\angle ACB \leq 45°$, triangle $\triangle AOB$ has circumradius

$$\frac{AB}{2\sin \angle AOB} = \frac{2R \sin \angle ACB}{2\sin(2\angle ACB)} = \frac{R}{2\cos \angle ACB} \leq \frac{R}{\sqrt{2}} < R,$$

and we contradict the minimality of R.

b) The triangle $\triangle ABC$ is acute. Note that $\angle BAC \geq 60°$. Now look at triangle $\triangle BOC$. The circumradius is

$$\frac{BC}{2\sin \angle BOC} = \frac{2R \sin \angle BAC}{2\sin(2\angle BAC)} = \frac{R}{2\cos \angle ABC} \leq \frac{R}{2\cos 60°} = R.$$

By minimality of R, we must have equality. Thus the triangle $\triangle ABC$ must be equilateral. However, in this case we get a contradiction with part a) since $\triangle BOC$ is an obtuse triangle with minimal circumradius R. □

Chapter 18

Solutions for Advanced Problems

1. Determine the number of 8×8 matrices in which each entry is a 0 or a 1 and each row and each column contains an odd number of 1's.

   Solution. We will build our matrix row by row. For a particular row of our matrix, we can choose either 0 or 1 for each of the first seven entries. The eigth entry, however, is determined by how many 1's appear amongst the first seven entries of the row: if there are already an odd number of 1's, the last element must be a 0; otherwise it will be a 1. Since there are 2 choices for each of the first 7 elements, there are 2^7 ways to construct such a row. This process works for each of the first 7 rows, giving $(2^7)^7 = 2^{49}$ ways. Now consider the final row. By similar logic to the final element of a row, this row is entirely predetermined since each element is the final element of a column. We can be assured that the final row will end up with an odd number of 1's by counting the parity of all 1's in the matrix. Explicitly, since we constructed the final row so that each column has an odd number of 1's, there must be an even number of 1's total in the matrix (the sum of eight odd numbers is even). There are an odd number of 1's in the first seven rows of the matrix (the sum of seven odd numbers is odd), thus there must be an odd number of 1's in the final row so that our overall count is even. Thus, there are 2^{49} matrices satisfying our given conditions. □

2. How many seven-digit integers have exactly three distinct digits?

   Solution. First, let us count the number of strings of length 7 of the letters $\{X, Y, Z\}$ such that each letter is used at least once. This is an easy Inclusion-Exclusion count. There are 3^7 strings of length 7, but this includes strings where one of the three letters is missing. For each letter, there are 2^7 strings where that letter is missing, hence $3 \cdot 2^7$ total. If we just subtract this, then we will have subtracted off the 3 strings with just one letter twice. Hence the correct count is $3^7 - 3 \cdot 2^7 + 3$. Notice that this is closely related to the desired count. If we replace each of X, Y, and Z with a distinct digit, then we will get a seven-digit string with exactly three distinct digits. There are two things we need to worry about. First, we do not want to allow leading zeroes so that we actually get a seven-digit integer, not just a string. Second, there is a symmetry in strings of the letters X, Y, Z. For any such string, we can permute the letters to get another such string. For example, the number 1123132 is represented by $XXZYXYZ$ with $X = 1$, $Y = 3$, $Z = 2$ and also by $ZZYXZXY$ with $X = 3$, $Y = 2$, $Z = 1$. We observe that since there are $3! = 6$ permutations of the letters X, Y, Z, we end up counting each string 6 times. Thus we divide our original string count by 6 to fix our second problem. Now we assign numeric values to our letters. We can replace the first letter of our string with any digit other than zero, giving us 9 choices. The second letter to appear can be any digit except the first one (9 choices), and the third letter will be one of the remaining digits (8 choices). Overall, this gives us $9 \cdot 9 \cdot 8 \cdot (3^7 - 3 \cdot 2^7 + 3)/6 = 195048$ seven-digit integers with exactly three distinct digits. □

3. An animal shelter has n cats and $3n$ dogs. Every cat hates exactly three dogs, and no two cats hate the same dog. Find a formula for the number of ways to assign each cat a dog kennel-mate that the cat doesn't hate.

   Solution. This is a variation on derangements (seen in Example 34). Suppose for some k cats ($0 \le k \le n$) we know that those cats are assigned a dog that they hate. There are 3 ways for each of those k cats to be assigned a dog they hate. For the remaining $n - k$ cats, there are $3n - k$ ways to assign a kennel-mate for the first cat, $3n - k - 1$ ways to assign a kennel-mate to the next cat, down to $3n - k - (n - k) + 1 = 2n + 1$ ways to assign the final cat a kennel-mate. Note that the product of these terms gives $(3n - k)!/(2n)!$. Then there are $3^k \cdot (3n - k)!/(2n)!$ ways to assign kennel-mates such that the chosen k cats are unhappy with their partners. Since there are $\binom{n}{k}$ ways to choose a group of cats to annoy, by the Principle of Inclusion-Exclusion overall there are

   $$\sum_{k=0}^{n} (-1)^k \binom{n}{k} 3^k \frac{(3n-k)!}{(2n)!}$$

   ways to assign kennel-mates such that no cat ends up disgruntled. □

4. There are two distinguishable flagpoles, and there are 19 flags of which 10 are identical blue flags and 9 are identical green flags. Let N be the number of distinguishable arrangements using all of the flags in which each flagpole has at least one flag and no two green flags on either pole are adjacent. Find the remainder when N is divided by 1000.

(2008 AIME II)

Solution. We first line up our 10 blue flags. We add a red flag to symbolize the split in our two flagpoles; there are 11 spots between or to either side of the 10 blue flags, so there are 11 possibilities for where to place the red flag. Next we place our green flags such that there is at most one green flag in each gap between the red and blue flags. This ensures that each pair of green flags is separated by at least one blue flag or, in the case that the red flag separates a pair of green flags, that the green flags in question are on different poles. There are 12 spots to place the 9 green flags in, so there are $\binom{12}{9} = 220$ ways to insert the green flags and thus a total of $11 \cdot 220 = 2420$ arrangements.

The result gives us a unique way to place the flags on the poles as follows: starting from the left we place flags from the bottom up onto the first pole. When we reach the red flag, we switch to loading the other pole. The only case we need to account for is the case where one pole has no flags on it; this occurs when the red flag is all the way to the left or all the way to the right of the other flags. There are 2 choices for which end the red flag is on, and $\binom{11}{9} = 55$ arrangements of the remaining flags, so our final count of arrangements is $N = 2420 - 2 \cdot 55 = 2310$. Thus the remainder upon division by 1000 is 310. □

5. Derive a closed form expression (without summations) for the number of lists of m 1s and n 0s that have k runs of 1s, where a *run* is a maximal consecutive string of identical values.

Solution. First, we split our m 1s into k runs. Each run must contain at least one 1, so we have $m-k$ additional 1s to distribute. We can think of this as a stars and bars problem: we have $m-k$ stars (representing 1s) and $k-1$ bars (indicating breaks between the runs). There are $\binom{m-1}{k-1}$ ways to arrange these.

Next, we place our 0s. We use $k-1$ 0's to split up our k runs of 1s. The remaining $n-k+1$ 0s can be distributed in any of the $k+1$ spaces between, before, or after the k runs. We can again think of a stars and bars situation where the 0s are stars and we have k bars separating the possible spaces. This gives $\binom{n+1}{k}$ ways to place the 0s relative to the 1s. Overall, this gives us $\binom{m-1}{k-1}\binom{n+1}{k}$ lists with k runs of 1s. □

6. At a conference for superheroes and supervillains, 5 pairs of heroes and villains are giving a panel where they will sit in a row. Of course, if any superhero sits next to his or her archnemesis, complete chaos will break out and ruin the convention. How many ways can you arrange the panelists so that the program proceeds smoothly?

    Solution. We will use both complementary counting and the Principle of Inclusion-Exclusion to help us out here. First, we count the total number of ways we could arrange the 10 people, which is 10!. Next, we subtract off the number of ways for at least one of the hero/villain pairs to be seated together. There are $\binom{5}{1} = 5$ ways to choose which pair sits together, 2 ways to choose which of the the pair sits to the left, and then 9! ways to arrange everyone (treating the pair as a single unit since they must be next to each other). This gives a total of $\binom{5}{1} \cdot 2 \cdot 9!$ ways. However, we have overcounted the number of ways at least two pairs could be seated together so we must subtract off $\binom{5}{2} \cdot 2^2 \cdot 8!$ (there are $\binom{5}{2}$ ways to choose which two pairs are seated next to each other, 2 ways to arrange each pair, and 8! ways to seat everyone). Continuing this pattern, we end up with the result

    $$10! - \binom{5}{1} \cdot 2 \cdot 9! + \binom{5}{2} \cdot 2^2 \cdot 8! - \binom{5}{3} \cdot 2^3 \cdot 7! + \binom{5}{4} \cdot 2^4 \cdot 6! - \binom{5}{5} \cdot 2^5 \cdot 5!$$

    □

7. Consider 3 sets X, Y, Z with $|X| = n, |Y| = m, |Z| = r$, and $Z \subset Y$. Denote by $s_{m,n,r}$ the number of functions $f : X \to Y$ for which $Z \subseteq f(X)$. Prove that:

    $$s_{m,n,r} = m^n - \binom{r}{1}(m-1)^n + \binom{r}{2}(m-2)^n - \cdots + (-1)^r(m-r)^n.$$

    Solution. We use complementary counting and the Principle of Inclusion-Exclusion to obtain $s_{m,n,r}$. We know that in total, there are m^n functions $f : X \to Y$. If $Z \nsubseteq f(X)$, then there must be some element $z \in Z$ such that there is no $x \in X$ with $f(x) = z$. Consider a subset S containing k elements of Z. The number of functions such that no element of S is in $f(X)$ is $(m-k)^n$ since there are $m-k$ remaining elements of Y for

Solutions for Advanced Problems 199

each element of X to be mapped to. There are $\binom{r}{k}$ subsets of Z of size k, so by the Principle of Inclusion-Exclusion we have

$$s_{m,n,r} = \sum_{k=0}^{r} (-1)^k \binom{r}{k} (m-k)^n$$

$$= m^n - \binom{r}{1}(m-1)^n + \binom{r}{2}(m-2)^n - \cdots + (-1)^r (m-r)^n. \quad \square$$

8. A collection of letters consists of n X's and r Y's. Find the number of different words (sequences) that can be formed from the X's and Y's if each sequence must contain n X's (and not necessarily all Y's).

Solution. This is a stars and bars type of problem only we have X's instead of bars and Y's instead of stars. We add an extra X and arrange the $n+1$ X's and r Y's. Our sequence will consist of everything appearing before the $n+1$st X; note that every sequence obtained this way is unique. This process ensures we use all n X's, but any Y's after the $(n+1)$-st X are discarded, so we do not necessarily have to use them all. We have $n+r+1$ characters in all and must choose r slots for the Y's; thus there are $\binom{n+r+1}{r}$ such sequences. $\quad \square$

9. Define an ordered triple (A, B, C) of sets to be *minimally intersecting* if $|A \cap B| = |B \cap C| = |C \cap A| = 1$ and $A \cap B \cap C = \emptyset$. For example, $(\{1,2\}, \{2,3\}, \{1,3,4\})$ is a minimally intersecting triple. Find the number of minimally intersecting ordered triples of sets for which each set is a subset of $\{1, 2, 3, 4, 5, 6, 7\}$.

(variation of 2010 AIME I # 7)

Solution. We first choose the elements that will be in $A \cap B, B \cap C$, and $A \cap C$ respectively. Since there are 7 elements in our set, we have 7 choices for the element in $A \cap B$, 6 choices for the element in $B \cap C$, and 5 choices for the element in $A \cap C$. Now we have met our quota that $|A \cap B| = |B \cap C| = |C \cap A| = 1$ and we have not violated the condition that $A \cap B \cap C = \emptyset$. For each remaining element, we have 4 choices since none of these elements can appear in multiple sets: either the element appears in one of the three sets A, B, C or it appears in none of them. Since there are 4 remaining elements, we have $7 \cdot 6 \cdot 5 \cdot 4^4 = 53760$ ordered triples of minimally intersecting sets. $\quad \square$

10. A *palindrome* on the alphabet $\{H, T\}$ is a sequence of H's and T's which reads the same from left to right as from right to left. Thus,

$HTH, HTTH, HTHTH$, and $HTHHTH$ are palindromes of lengths $3, 4, 5$, and 6 respectively. Let $P(n)$ denote the number of palindromes of length n on $\{H, T\}$. For how many values of n is $1000 < P(n) < 10000$?

Solution. We'll determine a general form for $P(n)$ for two different cases: the case where n is even and the case where n is odd. If n is even, then we have 2 choices for each of the first $n/2$ entries of our sequence. The latter $n/2$ entries will then be assigned in such a way that the resulting sequence is a palindrome (i.e., so that the kth element matches the $(n - k + 1)$-st element for $k = n/2 + 1$ to n). Thus there are $2^{n/2}$ palindromes of length n for n even. Similarly if n is odd we can assign the first $(n+1)/2$ characters as either H or T, but the remaining characters are then determined by the fact we must have a palindrome. This gives a total of $2^{(n+1)/2}$ palindromes of length n when n is odd.

Our goal is to determine when $1000 < P(n) < 10000$. Since we have determined $P(n)$ is always a power of 2, we look for the smallest power of 2 greater than 1000 and the largest power of 2 less than 10000. We have $1024 = 2^{10}$ so the smallest n for which $P(n) > 1000$ is $10 = (n+1)/2$ or $n = 19$. We also have $8192 = 2^{13}$ so the largest n for which $P(n) < 10000$ is $13 = n/2$ or $n = 26$. Thus $1000 < P(n) < 10000$ for $19 \leq n \leq 26$. We can conclude that there are 8 values of n such that $1000 < P(n) < 10000$. □

11. How many nonnegative integral solutions does the equation $n_1 + n_2 + \cdots + n_k \leq m$ have?

Solution. We would like to find a clever way to turn the inequality into an equality so that we can take advantage of what we already know. To do so, we add a slack variable s and change our inequality to the equation $n_1 + n_2 + \cdots + n_k + s = m$ where $n_1, \ldots, n_k$ and s are all nonnegative integers. Note that with s being a nonnegative integer, we now cover all the possible values $n_1 + n_2 + \cdots + n_k$ could equal from 0 up to m. As discussed in Example 21, the number of solutions to such an equation is $\binom{m+k}{m}$ since we have $k + 1$ variables and m units. □

12. How many nonnegative integer solutions are there to $x + y + z = 30$ such that none of x, y, z is divisible by 3?

Solution. Notice that in order for the sum of x, y, and z to be 30 (a multiple of 3) with none of x, y, z being a multiple of 3, all three numbers must have the same remainder upon division by 3. We will look at two cases: when the remainder of each number is 1 and when the remainder of each number is 2. If x, y, z all have remainder 1 upon division by 3, we can write $x = 3a + 1, y = 3b + 1, z = 3c + 1$. This gives us

$$x + y + z = 3a + 1 + 3b + 1 + 3c + 1 = 3(a + b + c) + 3 = 30$$

which simplifies to $3(a + b + c) = 27$ and then $a + b + c = 9$ where a, b, c are nonnegative integers. We've seen how to solve this type of problem before in Example 21 using stars and bars to get a result of $\binom{11}{2}$. Simliarly, if x, y, z all have remainder 2 upon division by 3, we can write $x = 3a + 2, y = 3b + 2, z = 3c + 2$. This gives us $a + b + c = 8$ after simplifying, which has $\binom{10}{2}$ solutions in nonnegative integers. Thus, overall, there are $\binom{11}{2} + \binom{10}{2} = 100$ nonnegative integer solutions to $x + y + z = 30$ such that none of x, y, z is divisible by 3. □

13. Find the number of triples (a, b, c) of positive integers such that each positive integer is a number from 1 to 9 and the product abc is divisible by 10.

Solution. In order for abc to be divisible by 10, at least one of a, b, c must be even and at least one must be divisible by 5. Since all three are integers from 1 to 9 this implies one of them must be exactly 5. We will use complementary counting along with the Principle of Inclusion-Exclusion here. First, we will count all possible ordered triples with no restrictions. Next, we will subtract off the triples with no 5 and the triples with no even numbers. At that point, however, we will have subtracted off those triples that have no even numbers and no 5 twice so we must add these back.

There are 9 choices for each of a, b, c so there are 9^3 total triples with no restrictions. If there are no 5's amongst the three numbers, there are only 8 choices remaining for each, so there are 8^3 triples that contain no 5's. If we do not allow even numbers, there are only 5 possible choices for each number $(1, 3, 5, 7, 9)$, so there are 5^3 triples containing no even numbers. Lastly, if we exclude both even numbers and 5, there are 4 choices remaining for each number $(1, 3, 7, 9)$ and thus 4^3 triples. This gives us as our final answer

$$9^3 - 8^3 - 5^3 + 4^3 = 156$$

ordered triples such that abc is divisible by 10. □

14. Find the number of ways of arranging the numbers $1, 2, \ldots, 8$ into three non-empty sets. For example, $\{1, 3, 6, 7\}, \{2, 5\}$, and $\{4, 8\}$ is one arrangement. The order of the sets does not matter.

Solution. First, suppose the order of the sets *does* matter. In this case, we have 3 choices for which set to put each of the 8 numbers in, so there are 3^8 arrangements of the numbers into three sets. However, this does not exclude the possibility that one or more of the sets are empty. If at least one of the sets is empty, there are $\binom{3}{1} = 3$ ways to choose which set must be empty and then 2 choices for each number which set to assign it to for a total of $3 \cdot 2^8$. However, we have now undercounted the case where two of the sets are empty. There are 3 ways for this to happen (we simply choose which set is nonempty). Finally, we adjust our count to ensure that order of the sets do not matter. There are 3! ways to permute the three non-empty sets; since we do not care about order we will divide our result by 3!. This gives us

$$\frac{3^8 - 3 \cdot 2^8 + 3}{3!} = 966$$

valid arrangements of the numbers $1, 2, \ldots, 8$ into three non-empty sets such that the order of the sets does not matter. $\square$

15. Let S_1 and S_2 represent two binary strings of length n. The *Hamming distance* between S_1 and S_2, denoted by $\mathcal{H}(S_1, S_2)$, is the number of positions in which S_1 and S_2 differ. For example, $\mathcal{H}(001011, 101001) = 2$. Given positive integers n and k with $k \leq n$, count the number of ordered pairs (S_1, S_2) of two binary strings S_1 and S_2, each of length n, such that $\mathcal{H}(S_1, S_2) = k$.

    Solution. First we construct S_1. We have two choices for each of the n positions of S_1 (either 0 or 1) so there are 2^n ways to construct S_1. Next, we must choose which k positions of S_2 will differ from S_1. There are $\binom{n}{k}$ ways to select which positions will differ; once we have selected these positions S_2 is completely determined. Thus, there are $\binom{n}{k}2^n$ ordered pairs of binary strings with Hamming distance k. $\square$

16. How many 15-letter arrangements of 5 A's, 5 B's, and 5 C's have no A's in the first 5 letters, no B's in the next 5 letters, and no C's in the last 5 letters?

    (variation on 2003 AMC 12A)

    Solution. Suppose we use k B's in the first 5 letters (where $0 \leq k \leq 5$). Then we use $5 - k$ C's in the first 5 letters. There are $\binom{5}{k}$ ways to choose which slots contain B's; once we have picked spaces for the B's, the remainder must be filled with C's. The remaining k C's must be used in the second group of 5 letters along with $5 - k$ A's, so again we

Solutions for Advanced Problems   203

have $\binom{5}{k}$ ways to arrange these letters. Lastly there are k A's and $5-k$ B's in the last 5 letters and thus $\binom{5}{k}$ ways to arrange the letters. Overall this gives us

$$\sum_{k=0}^{k} \binom{5}{k}^3 = 2252 \text{ arrangements.} \qquad \square$$

17. Consider numbers such that all digits of the number are different, the first digit is not zero, and the sum of the digits is 36. There are $N \times 7!$ such numbers. What is the value of N?

    Solution. The sum of all the digits $0, 1, \ldots, 9$ is 45, so we must not use some group of digits whose sum is 9. We'll look at a few different cases based on how many digits we do not use. If we omit the digit 9, there are 9! ways to arrange the digits $0, 1, 2, \ldots, 8$. If zero ends up as the first digit, we remove it to obtain a number with digit sum 36 that does not contain 0 or 9.

    Another possibility is that we could leave out a pair of nonzero digits summing to 9. There are 4 such pairs (1 and 8, 2 and 7, 3 and 6, and 4 and 5). For each pair, there are 8! ways to arrange the remaining 8 digits; again if zero is the first digit, we remove it to end up with another valid number.

    Lastly, there is the possibility we leave out a triple of nonzero digits summing to 9. There are 3 such triples ($\{1, 2, 6\}, \{1, 3, 5\}, \{2, 3, 4\}$). For each triple, we have 7! ways to arrange the remaining 7 digits (removing zero if it is the first digit). We cannot remove four or more nonzero digits and still obtain a digit sum of 36 since $1+2+3+4 = 10$ and these are the smallest digits we could possibly remove. Thus overall we have $9! + 4 \cdot 8! + 3 \cdot 7! = (9 \cdot 8 + 4 \cdot 8 + 3) \cdot 7! = 107 \cdot 7!$ possible numbers so $N = 107$. $\qquad \square$

18. In a sequence of coin tosses one can keep a record of instances in which a tail is immediately followed by a head, a head is immediately followed by a head, etc. We denote these by TH, HH, etc. For example, in the sequence $HHTTHHHHTHHTTTT$ of 15 coin tosses we observe that there are five HH, three HT, two TH, and four TT subsequences. How many different sequences of 15 coin tosses will contain exactly two HH, three HT, four TH, and five TT subsequences?

    (1986 AIME)

    Solution. We first find a general structure for our sequence by focusing on the transitions from tails to heads and vice versa. Since we have three

HT and four TH, our sequence must take the form $T_1H_1T_2H_2T_3H_3T_4H_4$ where each T_i represents a string of tails (containing at least one tails) and H_i represents a string of heads (containing at least one heads). Since each string must contain at least one heads/tails, we must add two additional H's to some combination of the H_i and five T's to some combination of the T_i to obtain two HH and five TT. We know distributing two identical objects among four groups can be done in $\binom{5}{2} = 10$ ways and distributing five identical objects among four groups can be done in $\binom{8}{5} = 56$ ways, so there are $10 \cdot 56 = 560$ total sequences with the given set of subsequences. $\square$

19. The expression $(x+y+z)^{2006} + (x-y-z)^{2006}$ is simplified by expanding it and combining like terms. How many terms are in the simplified expression?

(2006 AMC 12A)

Solution. By the multinomial theorem we have

$$(x+y+z)^{2006} + (x-y-z)^{2006}$$
$$= \sum_{a+b+c=2006} \binom{2006}{a,b,c} x^a y^b z^c + \sum_{a+b+c=2006} \binom{2006}{a,b,c} x^a (-y)^b (-z)^c$$
$$= \sum_{a+b+c=2006} (1+(-1)^{b+c}) \binom{2006}{a,b,c} x^a y^b z^c.$$

The number of terms is the number of ordered triples of nonnegative integers (a,b,c) such that $a+b+c = 2006$ and $b+c$ is even (in which case $(-1)^{b+c} = 1$ and we have a nonzero coefficient). In order for $b+c$ to be even, a must be even. For a particular $a = 2k$, we have $2006 - 2k + 1$ terms (since b can range from 0 to $2006 - 2k$). Thus our total number of terms is

$$\sum_{k=0}^{1003} (2006 - 2k + 1) = 1004 \cdot 2007 - 2 \sum_{k=0}^{1003} k$$
$$= 1004 \cdot 2007 - 2 \frac{1004 \cdot 1003}{2} = 1004^2 = 1008016. \quad \square$$

20. The polynomial $1 - x + x^2 - x^3 + \cdots + x^{16} - x^{17}$ may be written in the form $a_0 + a_1 y + a_2 y^2 + \cdots + a_{16} y^{16} + a_{17} y^{17}$, where $y = x + 1$ and the a_i's are constants. Find the value of a_2.

(1986 AIME)

Solution. Since $y = x+1$, we know $x = y-1$. Substituting this into our polynomial, we have

$$1 - x + x^2 - x^3 + \cdots + x^{16} - x^{17}$$
$$= 1 - (y-1) + (y-1)^2 - (y-1)^3 + \cdots + (y-1)^{16} - (y-1)^{17}.$$

By the binomial theorem, we know that the coefficient of y^2 in $(y-1)^n$ is $(-1)^{n-2}\binom{n}{2}$, so the sum of the coefficients taken from above is

$$\binom{2}{2} + \binom{3}{2} + \cdots + \binom{16}{2} + \binom{17}{2}.$$

By the Hockey Stick identity, this is simply $\binom{18}{3} = 816$, so $a_2 = 816$.

Alternatively, we could observe that

$$1 - x + x^2 - x^3 + \cdots + x^{16} - x^{17} = \frac{1 - x^{18}}{1 + x} = \frac{1 - (y-1)^{18}}{y}$$

and thus the coefficient a_2 of y^2 is the coefficient of y^3 in $1 - (y-1)^{18}$. Applying the binomial theorem gives us

$$a_2 = -(-1)^{15}\binom{18}{3} = 816. \qquad \square$$

21. Let S be a set containing n elements. Prove that

$$\sum_{A \subseteq S} \sum_{B \subseteq S} |A \cap B| = n \cdot 4^{n-1}.$$

Solution. We offer a combinatorial proof. We will count the sum of the size of the intersection of two sets over all ordered pairs of subsets of S.

<u>Answer 1:</u> This is exactly what the left hand side counts. $|A \cap B|$ gives us the size of the intersection of A and B, and we allow A and B to range over all possible subsets of S.

<u>Answer 2:</u> Consider a particular element $x \in S$. We will determine how many times x is counted in the summation. The element x contributes 1 to $|A \cap B|$ when both A and B contain x. Thus we need to count how many ordered pairs of subsets (A, B) there are such that both A and B contain x. For each of the $n-1$ elements of S other than x, we have 4 options: the element is in A but not B, the element is in B but not A, the element is in neither A nor B, or the element is in both A and B. Thus there are 4^{n-1} ordered pairs of subsets (A, B) such that both

A and B contain x. Since this holds true for any $x \in S$ and $|S| = n$, the total sum is $n \cdot 4^{n-1}$.

Since our two expressions count the same thing, they must be equal. Thus we have
$$\sum_{A \subseteq S} \sum_{B \subseteq S} |A \cap B| = n \cdot 4^{n-1}$$
as desired. □

22. Show that any odd number not divisible by 5 must divide some number of the form 10101...01, an alternating string of 1's and 0's. For example, 13 divides 10101, 17 divides 101010101010101, 9 and 19 divide 10101010101010101.

Solution. Let n be an odd number not divisible by 5. Consider the first $n+1$ numbers of the given form:

$$1, 101, 10101, 1010101, \ldots, \underbrace{1010\cdots 101}_{2n+1 \text{ digits}}$$

There are n possible remainders upon division by n and we have $n+1$ numbers, so by the Pigeonhole Principle at least two of our numbers have the same remainder. Suppose these numbers are

$$\underbrace{1010\cdots 101}_{2i+1 \text{ digits}} \quad \text{and} \quad \underbrace{1010\cdots 101}_{2j+1 \text{ digits}} \quad \text{with } i > j$$

When we take the difference of our two numbers, we obtain a number divisible by n. We have

$$\underbrace{1010\cdots 101}_{2i+1 \text{ digits}} - \underbrace{1010\cdots 101}_{2j+1 \text{ digits}} = \underbrace{1010\cdots 101}_{2i-2j-1 \text{ digits}} \underbrace{00\cdots 0}_{2j+2 \text{ digits}} = \underbrace{1010\cdots 101}_{2i-2j-1 \text{ digits}} \cdot 10^{2j+2}.$$

Since we know n is odd and not divisible by 5, n shares no common factors with 10^{2j+2}. Thus, n must divide $\underbrace{1010\cdots 101}_{2i-2j-1 \text{ digits}}$, which is of the desired form. Thus every odd number not divisible by 5 divides some number of the form 10101...01. □

23. Show that for every 16-digit number there is a string of one or more consecutive digits such that the product of these digits is a perfect square.

(1991 Japan Mathematical Olympiad)

Solution. Suppose the kth digit of our 16-digit number is n_k (for $1 \leq k \leq 16$). Define $N_k = n_1 \cdot n_2 \cdot n_3 \cdots n_k$ (for $0 \leq k \leq 16$). So $N_0 = 1$, $N_1 = n_1$, $N_2 = n_1 \cdot n_2$, $N_3 = n_1 \cdot n_2 \cdot n_3$, and so on. Note that each N_k is a product of consecutive digits of our 16-digit number.

Next, notice that because each n_k is a digit, the prime factorizations of the N_k are of the form $2^{\alpha_1} 3^{\alpha_2} 5^{\alpha_3} 7^{\alpha_4}$ with $\alpha_i \geq 0$. There are $2^4 = 16$ different parity distributions for the 17 N_k so by the Pigeonhole Principle, at least two of these have the same parity distribution. Suppose these are N_i and N_j with $i < j$. Then

$$\frac{N_j}{N_i} = \frac{n_1 \cdot n_2 \cdots n_j}{n_1 \cdots n_i} = n_{i+1} \cdot n_{i+2} \cdots n_j$$

gives us a perfect square since dividing N_j by N_i results in us subtracting exponents of the same parity, resulting in all even exponents. Since this is in turn a product of consecutive digits of our original 16-digit number, we have proven the existence of a string of one or more consecutive digits such that the product of these digits is a perfect square. $\square$

24. 10 integers are chosen from 1 to 100. Prove that we can find 2 disjoint, non-empty subsets of the chosen integers such that the 2 subsets give the same sum of elements.

    Solution. Consider a set S consisting of 10 integers from 1 to 100. We know that there are $2^{10} - 1 = 1023$ non-empty subsets of S, and we also know that the sum of the elements of S is at most $\sum_{i=91}^{100} i = 955$. Note that any subset of S will have a sum of elements less than or equal to the element sum of S itself. By the Pigeonhole Principle since we have 1023 non-empty subsets of S and less that 955 possible element sums for these subsets, there must exist at least two subsets (say S_1 and S_2) of S with the same sum of elements. If S_1 and S_2 are disjoint, we are done. Otherwise, $S_1 \setminus (S_1 \cap S_2)$ and $S_2 \setminus (S_1 \cap S_2)$ will be disjoint, and since we are removing identical elements from each they will also have the same sum of their elements. Thus, for any set of 10 integers chosen from 1 to 100, we can find 2 disjoint, non-empty subsets of the chosen integers such that the 2 subsets give the same sum of elements, as desired. $\square$

25. Every point in a plane is either red, green, or blue. Prove that there exists a rectangle in the plane such that all of its vertices are the same color.

(USAMTS Year 18)

Solution. We will focus on the set of points in the plane (x, y) such that x and y are integers, $1 \leq x \leq 4$, and $1 \leq y \leq 82$. We will refer to the four points in our set with the same y value as a "row." The number of ways to color a particular row with red, green, or blue is $3^4 = 81$. Since our set has 82 rows in total, by the Pigeonhole Principle there exist two rows that are colored identically. Pick two such rows. Since there are 4 points in each row and 3 colors, by the Pigeonhole Principle some color appears at least twice in our chosen rows. Taking two points of the same color from one of our rows and the two points in the corresponding spots in our other row gives us a rectangle in the plane with all vertices the same color. $\square$

26. Jenny starts with a pile of n stones, where $n \geq 2$ is a positive integer. At each step, she takes a pile of stones and splits it into two smaller piles. If the two new piles have a stones and b stones, then she writes the product ab on a blackboard. She keeps repeating these steps, until each pile has exactly one stone. Prove that no matter how Jenny splits the stones, the sum of the numbers on the blackboard at the end is always the same.

(For example, if Jenny starts with a pile of 12 stones, she can split it into a pile of 5 stones and a pile of 7 stones, and writes the number $5 \cdot 7 = 35$ on the blackboard. She can then split the pile of 5 stones into a pile of 2 stones and a pile of 3 stones, and writes the number $2 \cdot 3 = 6$ on the blackboard.)

Solution. We first look at the results for some small n.

- ($n = 2$) If we start with a pile of 2 stones, the only possible way to split it is into two piles of 1 stone each. We write the number $1 \cdot 1 = 1$ on the board, and this is our final sum.

- ($n = 3$) Starting with a pile of 3 stones, we must split into a pile of 2 stones and a pile of 1 stone. We write $2 \cdot 1 = 2$ on the board. Then we split the pile of 2 into two piles with one stone each and write $1 \cdot 1 = 1$ on the board to give us a total sum of $2 + 1 = 3$.

- ($n = 4$) We have two possible ways to split a pile of 4 stones. Though the problem statement implies that either way our sum should come out the same, we will check both possibilities to confirm this is the case.

  We could split the pile of 4 into two piles of 2. Then we would split each of these in turn into two piles of 1. Our final sum will be $2 \cdot 2 + 1 \cdot 1 + 1 \cdot 1 = 6$.

Solutions for Advanced Problems 209

Alternatively, we could split the pile into one of 3 stones and one with 1 stone. The 3 stones pile must then be split as we saw before in the $n = 3$ case, so our sum is $3 \cdot 1 + 2 \cdot 1 + 1 \cdot 1 = 6$, which matches with our other split pattern.

- ($n = 5$) The sum when we start with a pile of 5 stones will end up being 10 (you may want to try some possible splits to convince yourself this is the case).

So far our sums are $1, 3, 6$, and 10. We notice that these are triangular numbers, so we conjecture that for a pile of n stones, our final sum will be $\frac{n(n-1)}{2}$. Now we need to prove that this is in fact the case no matter how we split the pile. We will do this by strong induction on n.

Base Case: We have already verified our formula holds for $n = 2$ above.

Induction Hypothesis: Suppose that for any pile of n stones with $2 \leq n \leq k$, our final sum is $\frac{n(n-1)}{2}$ no matter how we split our piles.

Inductive Step: Consider a pile of $k + 1$ stones. We must make some first split into two piles - say these piles have a and b stones respectively (with $a + b = k + 1$, $1 \leq a, b \leq k$). This contributes $a \cdot b$ to our final sum.

Next we need to split the pile of a stones into piles of 1 stone each. Since $1 \leq a \leq k$, we can apply our strong induction hypothesis and know that splitting this pile will contribute $\frac{a(a-1)}{2}$ to our final sum. Similarly dividing up the pile of b stones will add $\frac{b(b-1)}{2}$ to our sum. Thus overall we have

$$ab + \frac{a(a-1) + b(b-1)}{2} = \frac{a^2 + b^2 + 2ab - a - b}{2}$$
$$= \frac{(a+b-1)(a+b)}{2} = \frac{(k+1)k}{2}$$

as desired. By the principle of mathematical induction, this concludes our proof. □

27. Prove that $\chi(T; k) = k(k-1)^{n-1}$ for all trees T on n vertices.

Solution. We proceed by induction on n.

Base Case: ($n = 1$) We have a single vertex, and we can use any of the k colors, giving us k ways to color this tree. This agrees with $k(k-1)^0$, so our base case holds.

Induction Hypothesis: Suppose that for some $m \geq 1$ we have

$$\chi(T; k) = k(k-1)^{m-1}$$

for all trees T on m vertices.

Inductive Step: Consider a tree T on $n = m+1$ vertices. Since $m+1 \geq 2$, we know by Example 83 that T has at least two leaves. Suppose one leaf is ℓ. We also know by Example 84 that $T - \ell$ is a tree on m vertices. By our induction hypothesis, the number of ways to color $T - \ell$ with k colors is $k(k-1)^{m-1}$. Now we must assign a color to ℓ. Since ℓ is a leaf, it has exactly one neighbor; ℓ may not have the same color as its neighbor, but it can have any of the other $k - 1$ colors. Thus, by the Product Rule we have

$$\chi(T; k) = (k-1)\chi(T - \ell; k) = k(k-1)^m$$

as desired. By the principle of mathematical induction, this concludes our proof. □

28. Show that every graph on n vertices with at least n edges contains a cycle.

Solution. Let $G = (V, E)$ be a graph on n vertices with at least n edges. Suppose G has k-components, which we label $H_1, \ldots, H_k$, such that H_i contains n_i vertices. Then we must have $\sum_{i=1}^{k} n_i = n$. We claim there exists some component H_j with at least n_j edges. If this were not the case, each component H_i must contain fewer than n_i edges, implying the total number of edges in our graph is $|E| < \sum_{i=1}^{k} n_i = n$, a contradiction. Thus there must exist some connected component H_j with at least n_j edges. We claim that this component H_j contains a cycle. We know by Example 85 that if a graph on n vertices is a tree, then it must have exactly $n - 1$ edges. Since H_j has n_j vertices and at least n_j edges, this implies it cannot be a tree. The definition of a tree is a graph that is connected and acyclic; since H_j is by construction connected, this means our other requirement must be violated, that is H_j is not acyclic. This implies H_j, and thus the whole graph G, contains a cycle. □

29. Let $p > 2$ be a prime number.
Find the number of subsets of $\{1, 2, \ldots, p-1\}$ with sum divisible by p.

(Bulgaria, 2006)

Solution. We consider the generating function

$$\prod_{j=1}^{p-1}(1+X^j) = \sum_{A\subseteq\{1,2,\ldots,p-1\}} X^{\sigma(A)}$$

where by $\sigma(A)$ we denote the sum of the elements of A and by definition if we have the empty we take the sum to be zero. Now if we take $X = \omega$ a primitive p-th root of unity and we denote by a_i the number of sets A such that $\sigma(A) \equiv i \pmod{p}$ for $i = 0, 1, \ldots, p-1$ we get that

$$\prod_{j=1}^{p-1}(1+\omega^j) = a_0 + a_1\omega + \ldots + a_{p-1}\omega^{p-1}.$$

Now ω^j for $j = 1, 2, \ldots, p-1$ are roots of the polynomial

$$X^{p-1} + X^{p-2} + \ldots + X + 1 = \frac{X^p - 1}{X - 1} = \prod_{j=1}^{p-1}(X - \omega^j).$$

Plug in $X = -1$ to get $\prod_{j=1}^{p-1}(1+\omega^j) = 1$. Finally by what we argued in the lecture, Example 125, we can conclude that $a_0 - 1 = a_1 = a_2 = \ldots = a_{p-1}$ so that means $(p-1)(a_0 - 1) + a_0 = 2^{p-1}$ thus

$$a_0 = \frac{2^{p-1} + p - 1}{p}.$$ □

30. A deck of 32 cards has 2 different jokers each of which is numbered 0. There are 10 red cards numbered 1 through 10 and similarly for blue and green cards. One chooses a number of cards from the deck to form a hand. If a card in the hand is numbered k, then the value of the card is 2^k, and the value of the hand is sum of the values of the cards in hand. Determine the number of hands having the value 2004.

(CGMO 2004)

Solution. This is again a problem for which it is convenient to use a generating function. Namely, let

$$F(X) = (1+X)^2(1+X^2)^3(1+X^{2^2})^3 \ldots (1+X^{2^{10}})^3.$$

Then the desired number of hands is the coefficient of X^{2004} in $F(X)$. The first factor corresponds to the two jokers with value 2^0 and the remaining factors correspond to the three cards numbered k with a value of 2^k, $1 \le k \le 10$.

Again we are looking to rewrite F in a convenient way.

It is easy to see that $(1-X)^3(1+X)F(X) = (1-X^{2^{11}})^3$ by repeatedly using the identity $(1-a)(1+a) = 1 - a^2$. Thus

$$F(X) = \frac{(1-X^{2^{11}})^3}{(1-X)^3(1+X)}.$$

Replacing the numerator by 1 does not effect any coefficient below $X^{2^{11}} = X^{2048}$, so it suffices to find the coefficient of X^{2004} in $\frac{1}{(1-X)^3(1+X)}$.

As in Chapter 14, we have the decomposition

$$\frac{1}{(1-X)^3(1+X)} = \frac{1}{2(1-X)^3} + \frac{1}{4(1-X)^2} + \frac{1}{8(1-X)} + \frac{1}{8(1+X)}.$$

Next, we make use of the generalized binomial identity

$$(1+X)^\alpha = \sum_{n \ge 0} \binom{\alpha}{n} X^n$$

to obtain

$$\frac{1}{(1-X)^3(1+X)} = \sum_{n \ge 0} \left(\frac{1}{2}\binom{n+2}{2} + \frac{1}{4}\binom{n+1}{1} + \frac{1}{8} + \frac{(-1)^n}{8}\right) X^n.$$

Thus we obtain the result that there are

$$\frac{1}{2}\binom{2006}{2} + \frac{2005}{4} + \frac{1}{8} + \frac{1}{8} = \frac{1}{2}\binom{2006}{2} + \frac{2006}{4} = 1003^2 \text{ hands.} \qquad \square$$

31. The set of natural numbers is partitioned into finitely many arithmetic progressions $\{a_i + dr_i\}$, $1 \le i \le n$. Prove that:

(a) $\sum_{i=1}^{n} \frac{1}{r_i} = 1$.

(b) There exist $i \ne j$ such that $r_i = r_j$.

(c) $\sum_{i=1}^{n} \frac{a_i}{r_i} = \frac{n-1}{2}$.

Solution. Note that when we consider the geometric series of x, we have the following equality: $\sum_{n\geq 0} X^n = \sum_{i=1}^{n}\sum_{d\geq 0} X^{a_i+dr_i}$ Then

$$\frac{1}{1-X} = \sum_{i=1}^{n} X^{a_i} \sum_{d\geq 0}(X^{r_i})^d = \sum_{i=1}^{n} \frac{X^{a_i}}{1-X^{r_i}}. \quad (*)$$

It follows that

$$1 = \sum_{i=1}^{n} \frac{X^{a_i}(1-X)}{1-X^{r_i}} = \sum_{i=1}^{n} \frac{X^{a_i}}{1+X+X^2+...+X^{r_i-1}}.$$

For the first statement plug in in $X = 1$ and we get the desired result.

For the second part let us assume that all the numbers are distinct. Take r_1 to be the largest of them. Now if we plug in in $(*)$ a primitive root of unity of order r_1, say ω, of order r_1 we see that the left hand side makes sense so there should be some cancellation going on in the right hand side. That can only happen if there is a r_j with $\omega^{r_j} = 1$. But that would imply that $r_1 | r_j$, which contradicts the choice of r_1.

For the final part, we already have the identity

$$1 = \sum_{i=1}^{n} \frac{X^{a_i}}{1+X+X^2+...+X^{r_i-1}}.$$

If we take derivative of this relation we obtain that

$$0 = \sum_{i=1}^{n} \frac{a_i X^{a_i-1}(1+X+X^2+...+X^{r_i-1}) - X^{a_i}(1+2X+3X^2+...+(r_i-1)X^{r_i-2})}{(1+X+X^2+...+X^{r_i-1})^2}.$$

Again if we plug in $X = 1$ into this last relation we have that

$$0 = \sum_{i=1}^{n} \frac{a_i r_i - (1+2+...+(r_i-1))}{r_i^2}.$$

We can rewrite this as $0 = \sum_{i=1}^{n} \frac{a_i}{r_i} - \sum_{i=1}^{n} \frac{r_i(r_i-1)}{2r_i^2}$ and thus

$$\sum_{i=1}^{n} \frac{a_i}{r_i} = \sum_{i=1}^{n}\left(\frac{1}{2} - \frac{1}{r_i}\right) = \frac{n}{2} - \frac{1}{2}\sum_{i=1}^{n}\frac{1}{r_i} = \frac{n-1}{2}$$

using again the proven statement in the introductory problems. □

32. Find all natural numbers n for which there exist two distinct sets of integers $\{a_1, a_2, \ldots, a_n\}$ and $\{b_1, b_2, \ldots, b_n\}$ such that the multisets

$$\{a_i + a_j | 1 \leq i < j \leq n\} \quad \text{and} \quad \{b_i + b_j | 1 \leq i < j \leq n\},$$

coincide.

(Erdös-Selfridge)

Solution. We will prove that only powers of 2 satisfy the required property.

Consider $f(X) = \sum_{i=1}^{n} X^{a_i}$ and $g(X) = \sum_{i=1}^{n} X^{b_i}$.

Let's note that

$$f^2(X) = \sum_{i=1}^{n} X^{2a_i} + 2 \sum_{1 \leq i < j \leq n} X^{a_i + a_j}$$

and

$$g(X) = \sum_{i=1}^{n} X^{2b_i} + 2 \sum_{1 \leq i < j \leq n} X^{b_i + b_j}.$$

Thus $f^2(X) - g^2(X) = \sum_{i=1}^{n} X^{2a_i} - \sum_{i=1}^{n} X^{2b_i} = f(X^2) - g(X^2)$

Now the useful observation is that $f(1) = g(1) = n$ so that means $X - 1 | f(X) - g(X)$. So let $f(X) - g(X) = (X - 1)^k h(X)$ with $h(1) \neq 0$.

Then we can rewrite our relation as

$$(f(X) - g(X))(f(X) + g(X)) = (X^2 - 1)^k h(X^2)$$

and this further is equivalent to

$$(X - 1)^k h(X)(f(X) + g(X)) = (X^2 - 1)^k h(X^2).$$

We can divide both sides by $(X - 1)^k$ and we are led to

$$h(X)(f(X) + g(X)) = (X + 1)^k h(X^2).$$

Plug in $X = 1$ to obtain $2nh(1) = 2^k h(1)$ and we can simplify by $h(1)$ and so $n = 2^{k-1}$.

The construction of these sets can be done using induction on j, where $n = 2^j$. For $j = 1$ we can take $A_1 = \{1, 4\}$ and $B = \{2, 3\}$.

Now we construct recursively

$$A_{j+1} = A_j \cup (2^{j+1} + B_j) \quad \text{and} \quad B_{j+1} = B_j \cup (2^{j+1} + A_j),$$

for $j \geq 1$. Here we use the notation $x + M = \{x + m | m \in M\}$. We leave it up to the reader to check the details. □

33. Determine with proof whether there exist a subset X of the nonnegative integers with the following property : for any integer n there is exactly one solution to $a + 2b = n$ with $a, b \in X$.

(Adapted from USAMO 1996)

Solution. Will take 0 to be an element of X. Let's consider

$$f(T) = \sum_{a \in X} T^a.$$

We have

$$f(T)f(T^2) = \sum_{a,b \in X} T^{a+2b}$$

and by hypothesis this is equal to $\sum_{i \geq 0} T^i$ since each nonnegative number should write in a unique way as $a + 2b$. Thus we have

$$f(T)f(T^2) = \frac{1}{1-T}.$$

The trick is now to write this relation for T^2, T^4, T^8 and so on.
We will obtain that

$$\frac{f(T)}{f(T^{2^{2i+1}})} = \frac{1-T^2}{1-T} \cdot \frac{1-T^8}{1-T^4} \cdot \ldots \cdot \frac{1-T^{2^{2i+1}}}{1-T^{2^{2i}}} = \prod_{k=1}^{i}(1+T^{2^{2k+1}})$$

Letting i arbitrarily large and noting that 0 is in our set it is easy to see that $f(T^{2^{2i+1}})$ will become 1. Thus

$$f(T) = \prod_{k=1}^{\infty}(1+T^{2^{2k+1}}).$$

Putting in words our set will be made of natural numbers that have 0's on odd positions of their binary expansion, reading the expansion from right to left. □

34. Let n be a positive integer, and $X = \{1, 2, \ldots, 2n\}$. How many subsets S of X are there, such that no two elements $x, y \in S$ differ by 2?

Solution. For any such set S, let
$$A = \{a : 2a \in S\} \quad \text{and} \quad B = \{b : 2b - 1 \in S\}.$$

Then A and B are two subsets of $\{1, 2, \ldots, n\}$ such that no two elements differ by 1. From Example 72, we know that there are F_{n+2} possibilities for such a set. Conversely, given two such sets A and B, we can build a set S by taking $S = (2A) \cup (2B - 1)$. Thus there are F_{n+2}^2 such sets. $\square$

35. A rectangle is divided into 200×3 unit squares. Prove that the number of ways of splitting this rectangle into rectangles of size 1×2 is divisible by 3.

(Baltic Way 2005)

Solution. We will derive a recurrence relation for the number t_n of tillings of a $2n \times 3$ rectangle.

Let's see what can happen when we look at the first 2×3 rectangle in a tiling of a $(2n + 2) \times 3$ rectangle. We tile this 2×3 rectangle in 3 ways to reduce to a tiling of a $2n \times 3$ rectangle, but note that there are two extra possibilities.

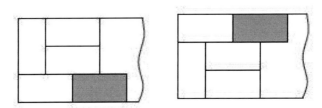

When this happens what remains is a tiling of a $(2n - 1) \times 3$ rectangle with one corner missing. Call the number of ways in which this can be done r_n. Then we have the recurrence $t_{n+1} = 3t_n + 2r_n$.

Now for the $(2n+1) \times 3$ rectangle with one corner missing there are only two ways to cover the 2×1 part at the end. We can use a 1×2 tile vertically. This leaves a $2n \times 3$ rectangle to be tiled, so we can do so in t_n ways. Alternately, we can use three horizontal 1×2 rectangles. This leaves a $(2n - 1) \times 3$ rectangle with one corner missing to be tiled, so we can do so in r_n ways.

Thus $r_{n+1} = t_n + r_n$. By the above $r_n = \dfrac{t_{n+1} - 3t_n}{2}$ so substituting into this we have the recurrence relation $t_{n+2} = 4t_{n+1} - t_n$.

We could use the initial values $t_1 = 3$ and $t_2 = 3t_1 + 2r_1 = 3 \cdot 3 + 2 \cdot 1 = 11$ and this recursion to obtain a general formula for t_n. However, since we only care about the remainder modulo 3, it is easier to step the recursion modulo 3 and see what happens. If we do this we find that $t_n \pmod 3$ is $0, 2, 2, 0, 1, 1, 0, 2, 2 \ldots$. Thus we see that t_n is a multiple of 3 precisely when $n = 3m + 1$. In particular, t_{100} is a multiple of 3. □

36. A word is a sequence of n letters of the alphabet $\{a, b, c, d\}$. A word is said to be complicated if it contains two consecutive groups of identic letters. The words *caab*, *baba* and *cababdc*, for example, are complicated words, while *bacba* and *dcbdc* are not. A word that is not complicated is a simple word. Prove that the numbers of simple words with n letters is greater than 2^n, if n is a positive integer.

(Romania TST 2003)

Solution. Let us denote by $S(n)$ the set of simple words of length n and by s_n the number of such words. If we add an extra letter at the end of each word from $S(n)$ we obtain a set $T(n+1)$ of words with $n+1$ letters and $|T(n+1)| = 4s_n$.

It is clear that $S_{(}n + 1) \subset T(n + 1)$, but they are not equal because adding the extra letter might make it complicated. Let's analyze when that happens.

Note that this happens when there are precisely two blocks of k letters repeating at the end, for $1 \leq k \leq m = \left\lfloor \dfrac{n+1}{2} \right\rfloor$.

Let $T_k(n+1)$ denote the set of these bad words with two blocks of length k repeating. We obviously have

$$s_{n+1} \geq |T(n+1)| - |T_1(n+1)| - |T_2(n+1)| - \ldots - |T_m(n+1)|$$

Now $|T_1(n+1)| = s_n$, and $|T_k(n+1)| \leq s_{n+1-k}$, because for such a word we can trim the last k letters and obtain a simple word with $n + 1 - k$ letter and obviously they are distinct.

Thus we have the inequality $s_{n+1} \geq 3s_n - s_{n-1} - s_{n-2} \ldots - s_{n+1-m}$.

It suffices to prove that $s_{n+1} > 2s_n$ by strong induction since $s_1 = 4$ and $s_2 = 12$.

The induction step is easy because using the above

$$s_{n+1} \geq 3s_n - \frac{1}{2}s_n - \frac{1}{4}s_n - \ldots > 3s_n - s_n = 2s_n. \qquad \Box$$

37. A permutation $\sigma : \{1, 2, \ldots, n\} \to \{1, 2, \ldots, n\}$ is called straight if and only if for each integer k, $1 \leq k \leq n-1$ the following inequality is fulfilled
$$|\sigma(k) - \sigma(k-1)| \leq 2$$

Find the smallest integer n for which there exist at least 2003 straight permutations.

(Romania TST 2003)

Solution. The main idea is to look where n is positioned. Following that idea, let us denote by x_n the number of all the straight permutations and by a_n the number of straight permutations having n in the first position, i.e., $\sigma(1) = n$. We will first find a recursion for a_n.

One can easily compute the first few terms $a_1 = 1$, $a_2 = 1$, $a_3 = 2$. Suppose that $n \geq 4$. We have two possibilities for the second position, $n-1$ and $n-2$. If $\sigma(2) = n-1$, then after deleting the initial n, we are left with a straight permutation of $\{1, 2, \ldots, n-1\}$ with $n-1$ in the first position, and conversely. Hence there are a_{n-1} such permutations.

The other possibility is that in the second position we have $n-2$, $\sigma(2) = n-2$. If $n-1$ is not in the third position, then it has only one possible neighbor $n-3$. Hence it must be in the last position. In this case, we have $\sigma(n) = n-1$ and $\sigma(n-1) = n-3$. Iterating this, we see $\sigma(3) = n-4$ and so on, thus there is only one permutation of this kind. On the other hand, if $\sigma(3) = n-1$, then it follows that $\sigma(4) = n-3$ and now deleting the first three terms gives a straight permutation of $\{1, 2, \ldots, n-3\}$, and conversely. Thus there are a_{n-3} straight permutations of this form. Combining all these cases, we get the recurrence:

$$a_n = a_{n-1} + a_{n-3} + 1.$$

Now let us examine the other positions where n could be. If n is in the last position, then by symmetry we again get a_n straight permutations. Otherwise, n is somewhere in the middle. Let $b_n = x_n - 2a_n$ for $n \geq 2$ be the number of such permutations. We now need a recurrence for b_n.

Note that if n is in the middle, then its two neighbors must be $n-1$ and $n-2$. If we erase the n, then the only new pair of neighbors is these two, hence we again get a straight permutation, with its two highest values adjacent. If $n-1$ is on the end, then this is the first case we analyzed in getting a recursion for a_n. For either end there are a_{n-2} such straight permutations, for a total of $2a_{n-2}$. If $n-1$ is not on either end, then it is in the middle (and so $n-2$ must be one of its neighbors). There are b_{n-1} of these by definition. Hence $b_n = b_{n-1} + 2a_{n-2}$.

One can turn this set of recursion into a recursion for just the sequence x_n. One possibility is $x_n = x_{n-1} + x_{n-3} + 2n - 4$, for $n \geq 5$. However, the characteristic polynomial of this recursion is the quartic $r^4 - r^3 - 1 = 0$ and the resulting formula for x_n would be unweildly. Instead, we can just use these recursions to calculate x_n. The terms grow relatively quickly, so this is not too hard. This calculation shows that the number we are looking for is $n = 16$.

n	a_n	b_n	x_n
1	1		1
2	1	0	2
3	2	2	6
4	4	4	12
5	6	8	20
6	9	16	34
7	14	28	56
8	21	46	88
9	31	74	136
10	46	116	208
11	68	178	314
12	100	270	470
13	147	406	700
14	216	606	1038
15	317	900	1534
16	465	1332	2262

□

38. 16 students took part in a mathematical competition where every problem was a multiple choice question with four choices. After the contest, it is found that any two students had at most one answer in common. Prove that there are at most 5 problems in the contest.

(China 1992)

Solution. Let's make a table of the students and problems, where the students are the columns and the problems are the rows and each cell we record the answer A, B, C or D.

We count again the number of agreements. By columns, we know from hypothesis, that any two students have at most one answer in common. Thus this number is at most the number of pairs of students, i.e $\leq \binom{16}{2} = 120$.

By rows, note that we can can just compute a lower bound for one of the rows and then multiply by n, the number of problems.

So let's look at problem 1 and say a of the students answered A, b of them answered B, c of them answered C and finally d of them answered D.

First we have $a + b + c + d = 16$. Now, the number of agreements is obviously $\binom{a}{2} + \binom{b}{2} + \binom{c}{2} + \binom{d}{2}$.

To get the desired lower bound we make us again of Jensen's inequality to see that $\binom{a}{2} + \binom{b}{2} + \binom{c}{2} + \binom{d}{2} \geq 4\binom{4}{2} = 24$.

Thus by rows we have at least $24n$ agreements on answers.

It follows, using the upper bound by columns, that $24n \leq 120$ and so $n \leq 5$. □

39. A group of 10 people went to a bookstore. It is known that everyone bought exactly 3 books and for every two persons, there is at least one book both of them bought. What is the least number of people that could have bought the book purchased by the greatest number of people?

(China 1993)

Solution. We draw again the usual table. Let's assume we have k books bought in total. Put the 10 people for the columns and on the rows we have the books. Mark in a cell a 1 if the person bought the book. Let a_i be the number of 1's in row i for $1 \leq i \leq k$. The question asks us what is the lowest possible value for $\max_i a_i$.

Summing by rows and columns we know that $\sum_{i=1}^{k} a_i = 30$. Secondly, we can again count the pairs of 1's. By columns this is at least $\binom{10}{2}$ since there is a common book for every pair of people.

On the other hand by rows we have that this is $\sum_{i=1}^{k}\binom{a_i}{2}$.

Thus we have that the following inequality should hold

$$\sum_{i=1}^{k}\binom{a_i}{2} \geq \binom{10}{2}.$$

Expanding out we get the equivalent

$$\sum_{i=1}^{k} a_i^2 \geq 90 + \sum_{i=1}^{k} a_i = 90 + 30 = 120.$$

Now let's note that $\max_{i} a_i \geq 5$. Indeed, if we assume the contrary we have

$$\sum_{i=1}^{k} a_i^2 \leq 4\sum_{i=1}^{k} a_i = 120$$

and thus we should have equality, namely every number should be equal to 4. But 4 does not divide 30 and we are done.

Also we can make a model for the situation. Take $k = 6$ for the books. Then we can assign the following:

$$B_1 \to \{P_1, P_2, P_3, P_5, P_6\}, \quad B_2 \to \{P_1, P_3, P_4, P_7, P_8\},$$
$$B_3 \to \{P_1, P_2, P_4, P_9, P_{10}\}, \quad B_4 \to \{P_2, P_5, P_6, P_7, P_8\},$$
$$B_5 \to \{P_3, P_7, P_8, P_9, P_{10}\}, \quad B_6 \to \{P_4, P_5, P_6, P_9, P_{10}\}.$$

Here the arrow stands for is bought and P_j are the persons. $\square$

40. Let X be a set with n elements. Given $k > 2$ subsets of X, each with at least r elements, show that we can find two of them whose intersection has at least $r - \dfrac{nk}{4(k-1)}$ elements.

(Iberoamerican 2001)

Solution. Call the subsets A_1, A_2, ..., A_k. We can assume that

$$\bigcup_{i=1}^{k} A_i = X$$

otherwise we can obtain a better bound using the statement for the cardinality of the union which is less than n. We consider the following table with rows indexed by the sets $A_1, A_2, \ldots, A_k$ and columns indexed by the elements $x_1, x_2, \ldots, x_n$, the elements of X. We put a 1 in the cell (i,j) with $1 \le i \le k$ and $1 \le j \le n$ if x_j is an element of the set A_i. Let a_i the number of 1's in the column i, or equivalently a_i is the number of sets in which the element x_i appears with $1 \le i \le n$. Now we have that

$$S = \sum_{i=1}^{n} a_i$$

is equal to the sum of the numbers in the whole table, and doing the sum by rows we obtain that

$$\sum_{i=1}^{n} a_i = \sum_{i=1}^{k} |A_i| \ge kr.$$

Let us look further at

$$\sum_{1 \le i < j \le k} |A_i \cap A_j|.$$

This is equal to the number of pairs of 1's which are in the same column. But this is equivalent in our notation to the fact

$$\sum_{1 \le i < j \le k} |A_i \cap A_j| = \sum_{i=1}^{n} \binom{a_i}{2}.$$

Using Jensen's inequality for the convex function $\binom{x}{2}$ we obtain that

$$\sum_{i=1}^{n} \binom{a_i}{2} \ge \frac{kr(kr-n)}{n}.$$

Thus

$$\sum_{1 \le i < j \le k} |A_i \cap A_j| \ge \frac{kr(kr-n)}{2n}$$

and thus the maximum of these intersection sizes, call it M, satisfies

$$M \ge \frac{r(kr-n)}{(k-1)n}.$$

But that doesn't look like what we want. Fortunately magic happens. We have that

$$r - \frac{nk}{4(k-1)} \le \frac{r(kr-n)}{(k-1)n}.$$

This is since it is equivalent to

$$4kr^2 - 4nr + n^2k - 4n(k-1)r \ge 0$$

and further this is the same as $k(2r-n)^2 \ge 0$ and we are done. $\square$

41. Let X be a finite set with n elements and let A_1, A_2, ..., A_m be three element subsets of X such that $|A_i \cap A_j| \leq 1$ for all $i \neq j$. Show that there exists a subset A of X with at least $\lfloor \sqrt{2n} \rfloor$ elements containing none of the A_i's.

Solution. Let A be the subset of X, containing none of the A_i's that has the maximum number of elements. Let $k = |A|$.

The key is the fact that we chose A to be maximal. Let's look at elements outside of A, from X.

We know by maximality that for any such $x \in X$, $A \cup \{x\}$ does not satisfy the conditions of the problem. Thus there is an index $i(x)$ with $A_{i(x)} \subset A \cup \{x\}$.

We can also rewrite this as $x \in A_{i(x)}$ and $C_x = A_{i(x)} \setminus \{x\} \subset A$ with 2 elements. Now because $|A_i \cap A_j| \leq 1$ all these sets C_x are distinct.

These sets C_x are now 2 element subsets of A, so there are at most $\binom{k}{2}$.

Since for each x in $X - A$ we have a set, we obtain the inequality

$$n - k \leq \binom{k}{2}.$$

Expanding out this is the quadratic $k^2 + k - 2n \geq 0$. Thus

$$k \geq \frac{-1 + \sqrt{8n+1}}{2} > \sqrt{2n} - 1$$

and we are done. $\square$

42. Let T be a finite set of integers greater than 1. A subset S of T is *good* if for any $t \in T$ one can find $s \in S$ such that s and t are not relatively prime. Prove that the number of *good* subsets of T is odd.

(USA TST 2010)

Solution. For any subset S of T, call a **prim** of S to be an integer in T that is coprime to all elements of S. Observe that a good set does not have a **prim**. For any set S, let $P(S)$ denote the set of all prims of S. (So, if S is a good subset, $P(S) = \emptyset$.)

For any two subsets A, B of T, call the ordered pair (A, B) **bad** if and only if for any $x \in A$ and $y \in B$, $\gcd(x, y) = 1$.

It is easy to see that the number of these ordered pairs is odd, observing the relation is symmetric. If (A, B) is a **bad** pair, then (B, A) is a **bad** pair, and $(\emptyset, \emptyset)$ is the only pair of the form (A,A)).

Now, for any subset X of T, the number of bad pairs of the form (X, S) is equal to $2^{|P(X)|}$ because (X, S) is **bad** if and only if S is a subset of $P(X)$.

Since $2^{|P(X)|}$ is odd iff $|P(X)| = 0$, i. e., if and only if X is a good subset, the above two facts imply that the number of good subsets of T is odd. □

43. Prove that for any set of n points in the plane there at most $cn\sqrt{n}$ distances among these points equal to 1, for some absolute constant $c > 0$.

    Solution. Label the points P_i, for $1 \le i \le n$ and also construct C_i the circles of radius 1 centered at these points. Let a_i be the number of points on each of the n circles. We are interested in bounding $S = \frac{1}{2}\sum_{i=1}^{n} a_i$, since each distance equal to 1 is counted twice.

    Let us count in two ways the triplets (P_i, A, B) with A, B on the circle C_i and $1 \le i \le n$. Conditioned on P_i, the number of points (A, B) is $\binom{a_i}{2}$ so in total the number of triplets is equal to $\sum_{i=1}^{n} \binom{a_i}{2}$.

    On the other hand since $AP_i = 1$ and $BP_i = 1$ it follows that P_i must be on both the circles centered at A and B of radius 1. Since any two circles intersect in at most two points and the number of pairs (A, B) is $\binom{n}{2}$, it follows that the number of triplets is at most $n(n-1)$.

    Thus we obtain that
    $$\sum_{i=1}^{n} \binom{a_i}{2} \le n(n-1).$$

    The only thing left to do is use the convexity of the function $\binom{x}{2}$ and apply Jensen to obtain that
    $$\frac{S(2S-n)}{n^2} \le n(n-1).$$

    We have the quadratic inequality $2S^2 - nS - n^2(n-1) \le 0$ and thus
    $$S \le \frac{n + n\sqrt{8n-7}}{4}.$$
    □

44. The numbers from 1 through 2015 are written on a blackboard. Every second, Dr. Math erases four numbers of the form a, b, c, $a+b+c$, and replaces them with the numbers $a+b$, $b+c$, $c+a$. Prove that this can continue for at most 9 minutes.

Solution. The first thing to note is that the number of numbers written on the board decreases by 1 for every move.
Secondly, another easy observation is that S, the sum of the numbers on the board, is invariant under the operations.
The key is to note the identity
$$(a+b)^2 + (b+c)^2 + (a+c)^2 = a^2 + b^2 + c^2 + (a+b+c)^2,$$
which means that the sum of the squares of the numbers also stays invariant. Thus for a given n, say we have a_1, ..., a_n are the numbers on the board. Then we have
$$\sum_{i=1}^{n} a_i = 1008 \cdot 2015 \quad \text{and} \quad \sum_{i=1}^{n} a_i^2 = 336 \cdot 2015 \cdot 4031.$$
Thus applying the Cauchy Schwartz inequality
$$n \sum_{i=1}^{n} a_i^2 \geq \left(\sum_{i=1}^{n} a_i \right)^2$$
and so we must have $n \geq \dfrac{3 \cdot 1008 \cdot 2015}{4031} = 1511.6$.
This means that process can be iterated at most $2015 - 1512 = 503$ and thus it can continue for at most $\dfrac{503}{60} = 8.39$ minutes. $\square$

45. Several stones are placed on an infinite (in both directions) strip of squares, indexed by the integers. We perform a sequence of moves, each move being one of the following two types:

(a) Remove one stone from each of the squares $n-1$ and n and place one stone on square $n+1$.

(b) Remove two stones from square n and place one stone on each of the squares $n-2$ and $n+1$.

Prove that any sequence of such moves will lead to a position in which no further moves can be made, and moreover that this position is independent of the sequence of moves.

(Russia 1997)

Solution. Let r be the positive root of the equation $x^2 - x - 1 = 0$. (The reader may notice that r is the golden ratio, but one can solve the problem without noticing this.) For any configuration of stones S, consider the sum $I(S) = \sum_i s_i r^i$ where s_i is the number of stones on the square i.

For transformation (a), the net change in $I(S)$ is

$$r^{n+1} - r^n - r^{n-1} = r^{n-1}(r^2 - r - 1) = 0.$$

For transformation (b), the net change in $I(S)$ is

$$r^{n+1} - 2r^n + r^{n-2} = r^{n-2}(r-1)(r^2 - r - 1) = 0.$$

Thus we see that $I(S)$ is invariant under the transformations of the problem.

We will show by induction on n, the number of stones, that the sequence of moves must end in a finite time. Since $r > 1$ there is a positive integer M such $r^M > I(S)$. This means that no matter what operations we do, we can never place a stone past the square labeled with $M - 1$.

Thus eventually we will move a stone to a rightmost position and never move it from there. Throw away this stone and use the induction hypothesis.

Now suppose that we could have two distinct ending configurations, call them A and B, and let the number of stones on the square i for these configurations be a_i and b_i, respectively. Note that since these are ending configurations, neither can have two stones in the same square or stones on each of two consecutive squares. Now take k to be the maximal index with $a_k \neq b_k$ and assume without loss of generality that $a_k = 1 > b_k = 0$. Note that the next stone in configuration B is at most in square $k - 1$ and then we have gaps of at least 1 between consecutive stones in B. Throw away the stones on positions $k+1, k+2, \ldots$ and call the new configurations A' and B'. Then

$$I(A') \geq r^k = \frac{r^{k+1}}{r^2 - 1} = r^{k-1} + r^{k-3} + \ldots > I(B').$$

This is a contradiction since the quantity I is invariant under moves, so we are done. $\square$

46. Consider a matrix whose entries are integers. Adding the same integer to all entries on a row, or in a column, is called an operation. It is given that, for infinitely many positive integers n, one can obtain, through a

finite number of operations, a matrix having all entries divisible by n. Prove that, through a finite number of operations, one can obtain the null matrix.

(Romania 2009)

Solution. Consider any two rows and any two columns, and suppose their intersections form $\begin{pmatrix} a & b \\ c & d \end{pmatrix}$. Notice that if we do any operation, $a+d-b-c$ is invariant. When all entries are divisible by n, $a+d-b-c$ is also divisible by n. So $a + d - b - c$ is divisible by infinitely many integers and is thus 0. This applies for any pair of rows and columns.

Now add an appropriate amount to each row so that every entry in the first column is 0. Then add an appropriate amount to each column so that every entry in the first row is 0. For the matrix entry x in row $m > 1$, column $n > 1$, considering rows 1 and m and columns 1 and n gets the matrix $\begin{pmatrix} 0 & 0 \\ 0 & x \end{pmatrix}$. By the above paragraph, $x = 0$. So in fact the entire matrix is 0. $\square$

47. At the vertices of a regular hexagon six nonnegative integers are written whose sum is n. One is allowed to make moves of the following form: (s)he may pick a vertex and replace the number written there by the absolute value of the difference between the numbers written at the two neighboring vertices. Prove that if n is odd then one can make a sequence of moves, after which the number 0 appears at all six vertices.

(USAMO 2003)

Solution. Let's first outline a strategy. There are two semi-invariants that we can think of at first sight: the number of zeros and the maximum of the six numbers. The former is monotonic only if we choose our moves carefully, otherwise this number could both increase and decrease. The latter however is a monotonically nonincreasing quantity. The only instance in which this latter quantity cannot be reduced further is when we have the numbers $a, a, 0, a, a, 0$ at the vertices (in that order), a deadlock situation. So if we somehow make sure that this configuration is never reached then we can get to all zeros.

Call a configuration $\mathcal{C} = (a_1, a_2, \ldots, a_6)$ *odd* if the sum $\sum_{i=1}^{6} a_i$ is odd. We denote with m_k the move done at vertex k, i.e replacing a_k with $|a_{k+1} - a_k|$, where we work with indices modulo 6. Note that if $M(\mathcal{C})$ is the maximum of the number written at the vertices, then no operation increases this number. We call configuration $\mathcal{C}$ *strongly odd* if the numbers at the vertices alternate between even and odd. We construct an algorithm that in the end decreases the maximum so it will reach at one point the desired state of all zeroes in the vertices. Note that since the sum of the numbers initially is n, we initally have an odd umber of odd numbers among the vertices.

• Suppose $\mathcal{C}$ is *odd* but not *strongly odd*. If we have five odd numbers, we can assume
$$\mathcal{C} \equiv (1, 1, \ldots, 1, 0) \pmod{2}.$$

Applying m_2 and m_6 we will get a *strongly odd configuration*.
If we have three odd numbers but they do not alternate on the vertices, then using again symmetry we have two possibilities:
$$\mathcal{C} \equiv (1, 1, 1, 0, 0, 0) \pmod{2} \quad \text{or} \quad \mathcal{C} \equiv (1, 1, 0, 1, 0, 0) \pmod{2}.$$

In the first case apply t_4 and t_6 to reduce to five odd numbers, and for the latter apply t_6 and t_1 to get a *strongly odd* configuration.
If there is just one odd number, we can again by symmetry assume
$$\mathcal{C} \equiv (1, 0, 0, 0, 0, 0) \pmod{2}.$$

Apply t_2 and then t_6 to reduce to three odd numbers.

• Now suppose we have a *strongly odd* configuration
$$\mathcal{C} \equiv (1, 0, 1, 0, 1, 0) \pmod{2}.$$

Applying t_2, t_4 and t_6 we can assume that $M(\mathcal{C})$ is an odd number. If $M(\mathcal{C}) \neq a_1$, then applying t_3 and t_5 to get a odd configuration with $M(\mathcal{C}') < M(\mathcal{C})$ and then repeat step 1. A similar argument works for a_3 and a_5. Thus the last hop is when $M(\mathcal{C}) = a_1 = a_3 = a_5$. But then we can apply t_2, t_4 and t_6, followed by t_1, t_3, t_5 to get all the numbers equal to 0.

Alternate Solution. Let the six numbers be a_0, a_1, a_2, a_3, a_4, and a_5, where we treat indices as being mod 6. Since the sum of all 6 is odd, without loss of generality, we may assume $a_0 + a_2 + a_4$ is odd.

Claim. If a_0, a_2, a_4 are not all equal and at least two of them are nonzero, then we can do moves to decrease $\max(a_0, a_2, a_4)$ and keep the sum of these three entries odd.

Proof. Without loss, we may assume $a_0 \leq a_2 \leq a_4$ and hence $a_0 < a_4$ and $a_2 \neq 0$. After doing moves at positions 1, 3, and 5, the six numbers will be:
$$(a_0, a_2 - a_0, a_2, a_4 - a_2, a_4, a_4 - a_0).$$

If $a_0 > 0$, then do moves at positions 2 and 4. The result will be

$$(a_0, a_2 - a_0, a_4 - a_0, a_4 - a_2, a_2 - a_0, a_4 - a_0).$$

This decreases the maximum since a_4 is strictly larger than a_0, $a_4 - a_0$, and $a_2 - a_0$. The new sum of these three entries will be

$$a_0' + a_2' + a_4' = a_4 + a_2 - a_0 = (a_0 + a_2 + a_4) - 2a_0$$

which is still odd.

If $a_0 = 0$, then because the sum is odd, we must have $0 < a_2 < a_4$. Now do moves at positions 0 and 4. The result will be

$$(a_4 - a_2, a_2, a_2, a_4 - a_2, a_2, a_4).$$

This decreases $\max(a_0, a_2, a_4)$ but keeps the sum of these three unchanged, hence odd.

Iteratively applying this claim, we cannot decrease the maximum indefinitely. Hence we must eventually wind up with either $a_0 = a_2 = a_4 = a$ or two of these three equal to zero, say $a_0 = a_2 = 0$ and $a_4 = a$. In the first case, doing moves at positions 1, 3, and 5 will result in $(a, 0, a, 0, a, 0)$. Then doing moves at positions 0, 2, and 4 will give $(0, 0, 0, 0, 0, 0)$. In the second case, doing moves at positions 1, 3, and 5 will result in $(0, 0, 0, a, a, a)$. Doing a move at position 4 will give $(0, 0, 0, a, 0, a)$. Finally, doing moves at positions 3 and 5 will give $(0, 0, 0, 0, 0, 0)$. $\square$

48. A $(2n+1) \times (2n+1)$ board is going to be tiled with pieces of the types as shown

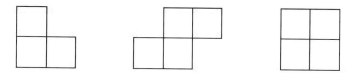

where rotations and reflections of the tiles are allowed.

Prove that at least $4n + 3$ pieces of the first type will be used.

(IMO Shortlist 2002)

Solution. Number the rows and columns from 1 to $2n+1$. Color in black the squares that are in an odd row and an odd column and color white the rest of the squares. There are $(n+1)^2 = n^2 + 2n + 1$ black squares and $3n^2 + 2n$ white squares. The pieces of the first type can cover two white squares and one black square or three white squares. The rest of the pieces cover always three white squares and one black square.

Let A be the number of pieces of the first kind that cover one black square, B the number of pieces of the first kind that cover no black squares and C the number of pieces of the other two kinds.

Then looking at black squares, since they are covered by pieces of type A and C we should have $A + C = n^2 + 2n + 1$.

Counting the white squares we have $2A + 3B + 3C = 3n^2 + 2n$. Multiplying the first relation by 3 and substracting the second one we get $A - 3B = 4n + 3$. But obviously $A - 3B \leq A + B$ so we get at least $4n + 3$ pieces of the first kind. □

49. A regular 2004-sided polygon is given, with all of its diagonals drawn. After some sides and diagonals are removed, every vertex has at most five segments coming out of it. Prove that one can color the vertices with two colors such that at least $\frac{3}{5}$ of the remaining segments have ends with different colors.

(MOP Homework 2004)

Solution. First note that any graph of maximum degree d can be properly vertex colored with d or $d+1$ colors. Thus our graph in the hypothesis can be colored with 6 colors.

Now choose 3 colors from the 6 uniformly at random (each occurs with the same probability) and collapse them into a single color-call it yellow. Do the same for the remaining 3 the same and call this color green. Now we have 2 coloring which is likely improper. The probability that any edge is monochromatic with respect to this new coloring is

$$\frac{2 \cdot 4}{\binom{6}{3}} = \frac{2}{5}$$

thus the expected number of monochromatic edges is at most $\frac{2|E|}{5}$. Thus there must be some particular choices for which no more than $\frac{2}{5}$ of the edges of G are monochromatic and thus at least $\frac{3}{5}$ of the edges will have endpoints of different colors. □

50. Let A be a set of N residues modulo N^2. Prove that there exists a set B of N residues modulo N^2 such that $A + B = \{a+b | a \in A, b \in B\}$ has at least $\frac{N^2}{2}$ elements.

(IMO Shortlist 1999)

Solution. Pick a random collection of n elements of $\mathbb{Z}/N^2\mathbb{Z}$, each of the n elements being taken with probability $\frac{1}{N^2}$ and all choices being independent. Put the distinct elements among the n chosen ones in a set B, which may have fewer than n elements. Consider the random variable $X = |A + B|$. Then if we denote by X_i the random variables taking value 1 if $i \in A + B$ and 0 otherwise, for $0 \le i \le N^2 - 1$ we have that $X = \sum_{i=0}^{N^2-1} X_i$.

Now using linearity of expectation we get that

$$E[X] = \sum_{i=0}^{n^2-1} E[X_i] = \sum_{i=0}^{N^2-1} P(i \in A + B)$$

The probability $P(i \notin A + B)$ is easy to count, since we can translate N times is just the n th power of the probability that an element is not in A so it is equal to $\left(1 - \frac{N}{N^2}\right)^N$, since A has N elements and so

$$P(i \in A + B) = 1 - \left(1 - \frac{1}{N}\right)^N$$

From the inequality $\left(1 - \frac{1}{N}\right)^N < \frac{1}{2}$, it follows that $E[X] > \frac{N^2}{2}$ and we are done. □

51. An $m \times n$ checkerboard is colored randomly: each square is independently assigned red or black with probability $\frac{1}{2}$. we say that two squares, p and q, are in the same connected monochromatic region if there is a sequence of squares, all of the same color, starting at p and ending at q, in which

successive squares in the sequence share a common side. Show that the expected number of connected monochromatic regions is greater than $\frac{mn}{8}$.

(Putnam 2004)

Solution. This solution is due to Noam Elkies.

Number the squares of the checkerboard 1, ..., mn by numbering the first row from left to right, then the second row, and so on. We prove by induction on i that if we just consider the figure formed by the first i squares, its expected number of monochromatic components is at least $\frac{i}{8}$.

For $i = 1$, this is clear. For the induction step suppose the i-th square does not abut the left edge or the top row of the board. Then we may divide into three cases

- With probability $\frac{1}{4}$, the i-th square is opposite in color from the adjacent squares directly above and to the left of it. In this case adding the i-th square adds one component.

- With probability $\frac{1}{8}$, the i-th square is the same in color as the adjacent squares directly above and to the left of it, but opposite in color from its diagonal neighbor above and to the left. In this case, adding the i-th square either removes a component or leaves the number unchanged.

- In all other cases, the number of components remains unchanged upon adding the i-th square.

Hence adding the i-th square increases the expected number of components by at least $\frac{1}{4} - \frac{1}{8} = \frac{1}{8}$ and we are done using the induction step.

If the i-th square does abut the left edge of the board, the situation is easier: if the i-th square differs in color from the square above it, one component is added, otherwise the number does not change. Hence adding the i-th square increases the expected number of components by $\frac{1}{2}$; likewise if the i-th square abuts the top edge of the board. Thus the expected number of components is at least $\frac{i}{8}$ by induction, as desired. $\square$

52. A finite collection of squares has total area 4. Show that they can be arranged to cover a square of side 1.

Solution. Let a_i for $1 \leq i \leq k$ be the sidelengths of the collection of squares we have. If one of the a_i's is bigger than 1 we are done. Otherwise, for each i we can find a positive integer n_i such that $a_i \in [2^{-n_i}; 2^{-n_i+1})$.

Covering the square will only get harder if we replace the given squares by squares of sidelengths $b_i = \dfrac{1}{2^{n_i}} \leq a_i$. Because $a_i < 2b_i$, these new squares will have a sum of areas greater than 1.

Let's show that these new squares can cover the unit square, and we will be done. The strategy is simple. Divide the square into 4 squares of side $\dfrac{1}{2}$. Use, if we have any, the squares in our new collection of side $\dfrac{1}{2}$. If some squares remain, divide these again in 4 squares of side $\dfrac{1}{4}$. We will cover all the squares and hence be done at some point with this procedure, because the sum of the areas of the covering squares is greater than 1. □

53. Given $2n+3$ points in the plane, no three collinear and no four on a circle, prove that there exists a circle containing three of the points such that exactly n of the remaining points are in its interior.

Solution. Let A and B be two consecutive points on the convex hull of the $2n+3$ points. By choosing a favorable coordinate system we may arrange that the line AB is horizontal and the other points, call them $C_1, C_2, \ldots, C_{2n+1}$, have larger y-coordinates.

Next let us look the collection of angles $\angle AC_iB$ for $1 \leq i \leq 2n+1$. These are all distinct by the hypothesis, since no four points can be on a circle.

Thus we can assume without loss of generality that $\angle AC_iB < \angle AC_{i+1}B$, for all $1 \leq i \leq 2n$. We claim that a good circle is the circumscribed circle of $\triangle AC_{n+1}B$.

To see this note that since all points are on the same side of AB, if $i > n+1$ then $\angle AC_iB > \angle AC_{n+1}B$, so C_i is inside of the circle. A similar argument shows that if $i < n+1$, then C_i is outside the circle. □

54. Let $n \geq 4$ be a fixed positive integer. Given a set $S = \{P_1, P_2, \ldots, P_n\}$ of n points in the plane such that no three are collinear and no four

concyclic, let a_t, $1 \leq t \leq n$, be the number of circles $P_i P_j P_k$ that contain P_t in their interior, and let $m(S) = \sum_{i=1}^{n} a_i$. Prove that there exists a positive integer $f(n)$, depending only on n, such that the points of S are the vertices of a convex polygon if and only if $m(S) = f(n)$.

(IMO Shortlist 2000)

Solution. We claim that the function $f(n) = 2\binom{n}{4}$ satisfies the requirements.

Let the **weight** $w(a, b, c, d)$ of the four points P_a, P_b, P_c and P_d be the number of points P_i where $i \in \{a, b, c, d\}$ such that P_i is properly contained in the circle passing through the remaining three points. Observe that
$$m(S) = \sum_{1 \leq a < b < c < d \leq n} w(a, b, c, d)$$

It is easy to see that we need to analyze two cases: when the quadrilateral determined by the 4 points is convex and when it is concave.

Claim. The weight of a convex quadrilateral is 2.

Proof. Let the vertices of the convex quadrilateral be denoted as A, B, C and D. We have that point D lies within the circumcircle of $\triangle ABC$ if and only if
$$\angle ABC > 180 - \angle ADC \quad \Leftrightarrow \quad \angle ABC + \angle ADC > 180$$

Therefore if D lies within the circumcircle of $\triangle ABC$, it also follows by symmetry that B lies within the circumcircle of $\triangle ADC$. If not, then since the sum of the interior angles of $ABCD$ is 360,
$$\angle ABC + \angle ADC < 180 \quad \Rightarrow \quad \angle BAD + \angle ACD > 180$$

Therefore A lies within the circumcircle of $\triangle BCD$ and C lies within the circumcircle of $\triangle ABD$. In both cases, the weight of $ABCD$ is equal to 2. □

Claim. The weight of a concave quadrilateral is 1.

Proof. Let $ABCD$ denote the concave quadrilateral. Without the loss of generality, let A be such that interior angle $\angle BAC > 180$. It follows that A lies in the interior of triangle $\triangle BCD$. Therefore, the circumcircle of $\triangle BCD$ contains A. However, none of the remaining three circles passing through three of the points A, B, C and D contain the remaining point. Hence the score of $ABCD$ is equal to 1. □

If the vertices of S form a convex n-gon, then each quadrilateral formed by four distinct points in S is a convex quadrilateral and therefore

$$m(S) = \sum_{1 \leq a < b < c < d \leq n} w(a,b,c,d) = 2\binom{n}{4}$$

If the vertices of S form a concave n-gon, then at least one of the quadrilaterals formed by four distinct points in S is a concave quadrilateral and therefore

$$m(S) = \sum_{1 \leq a < b < c < d \leq n} w(a,b,c,d) < 2\binom{n}{4}$$

The function $f(n) = 2\binom{n}{4}$ therefore satisfies that $m(S) = f(n)$ if and only if the points in S are the vertices of a convex polygon. □

55. Let S be a set of n points in the plane. No three points of S are collinear. Prove that there exists a set P containing $2n - 5$ points satisfying the following condition: In the interior of every triangle whose three vertices are elements of S lies a point that is an element of P.

(IMO Shortlist 1991)

Solution. Let's consider the points in a coordinate system $P_i(x_i; y_i)$ such that $x_1 < x_2 < \ldots < x_n$. Let d be half of the minimum distance between a point P_i and a line $P_j P_k$, where (i, j, k) goes through all triplets of distinct indices from $\{1, 2, \ldots, n\}$.

We define the set $\mathcal{P}$ of $2n - 4$ points:

$$\mathcal{P} = \{(x_i; y_i - d), (x_i; y_i - d) | i = 2, 3, \ldots, n - 1\}$$

Now consider any triangle $P_k P_l P_m$ where $k < l < m$. Then its interior has to contain one of the points $(x_l; y_l + d)$ or $(x_l; y_l - d)$ so $\mathcal{P}$ is a good set and is close to what we need. We just need to eliminate one point from $\mathcal{P}$.

The convex hull of the points contains at least three points, which must include P_1 and P_n. Assume that a third point is P_j. Clearly one of the points $(x_j; y_j \pm d)$ is outside the convex hull and we can eliminate it from $\mathcal{P}$. □

56. $\mathcal{A}$ is a closed polygon set so that for any two points in $\mathcal{A}$, the line segment joining the two points lies completely in $\mathcal{A}$. Prove that there exists a point O in $\mathcal{A}$, such that for any points X, X' on the boundary of $\mathcal{A}$, such that O lies on line segment XX' we have

$$\frac{1}{2} \le \frac{OX}{OX'} \le 2.$$

(Iran 2004)

Solution. For all M on the boundary of $\mathcal{A}$, let $\mathcal{A}_M$ be the homothetic of $\mathcal{A}$ with respect to the homothety h_M of pole M and ratio $\frac{2}{3}$. Using the convexity of $\mathcal{A}$, we deduce that $\mathcal{A}_M$ is convex and contained in $\mathcal{A}$ for all such point M.

Using the convexity of $\mathcal{A}$ and $\mathcal{A}_M$, we can see that if M, N, P are three points on the boundary of A and G is the center of gravity of MNP then G belongs to each of the sets $\mathcal{A}_M, \mathcal{A}_N, \mathcal{A}_P$. It follows that the family of the sets $\mathcal{A}_M$ is a set of bounded convex polygons such that any three have a common point.

From Helly's theorem, we deduce that there exists a point O which belongs to all the sets $\mathcal{A}_M$ where M is a point of the boundary of $\mathcal{A}$.

Now, let's consider any line l which passes through O. This line meet the boundary of $\mathcal{A}$ in two points X, X'. Then, from above O belongs to the segment XX_2 where $X_2 = h_X(X')$, and to the segment XX'. Thus, $OX \le XX_2 = \frac{2}{3}XX'$ so that $OX' \ge \frac{1}{3}XX'$. Therefore $\frac{OX}{OX'} \le 2$.

The other inequality is obtained by interchanging X and X'. □

Appendix: Recurrence Relations

Here we include some theoretical facts about solving recurrence relations, needed in the chapter with the same title. This is by no means an exhaustive treatment of the topic, but it should suffice for the recurrence relations encountered in the chapter and the proposed problems.

Definition. A linear recurrence relation for a sequence a_n is of the type

$$a_n = c_1 a_{n-1} + c_2 a_{n-2} + \ldots + c_k a_{n-k} + b_n$$

where $c_1, c_2, \ldots, c_k, b_n$ may be functions of n, but not of the a_i.

For example, the recurrence relation for Catalan numbers, call them C_n, obeys the recurrence relation

$$C_n = C_1 C_{n-1} + C_2 C_{n-2} + \ldots + C_{n-1} C_1$$

is not a linear recurrence relation.

On the other hand, the recurrence for the Fibonacci numbers, usually denoted with F_n, satisfies the linear recursion

$$F_n = F_{n-1} + F_{n-2}.$$

Definition. A linear recurrence relations is called homogeneous if $b_n = 0$.

Obviously the Fibonacci numbers have a linear recurrence relations which is homogeneous.

Definition. A linear recurrence relation has constant coefficients if all the c_i are independent of n.

For example, the derangement number D_n (permutations of the set $\{1, 2, \ldots, n\}$ without any fixed points) satisfy the recurrence relation

$$D_n = (n-1)(D_{n-1} + D_{n-2}).$$

The methods for solving recurrence relations that we shall present will cover only recurrence relations with constant coefficients.

Before we move to state these results, a few more words are needed. A recurrence relation only gives us what the next term of the sequence; it does not provide any information about how the sequence starts. To pin the sequence down we need to include values for the first k terms (see the definition of linear recurrence). As an example the Fibonacci sequence is determined by the initial values $F_0 = F_1 = 1$. On the other hand, the Lucas sequence obeys the same recurrence relations but it starts off with $L_0 = 2$ and $L_1 = 1$.

Definition. The characteristic polynomial of homogeneous recurrence relations with constant coefficients $a_n = c_1 a_{n-1} + c_2 a_{n-2} + \ldots + c_k a_{n-k}$ is

$$f(X) = X^k - c_1 X^{k-1} - \ldots - c_{k-1} X - c_k.$$

Theorem. *If the characteristic polynomial of the recurrence relations has distinct roots $z_1, \ldots, z_m$, such that the multiplicity of z_i is e_i, for $i = 1, \ldots, m$, then all the sequences having this characteristic polynomial can be written as linear combinations of*

$$z_1^n, \; nz_1^n, \ldots, \; n^{e_1-1} z_1^n;$$

$$z_2^n, \; nz_2^n, \ldots, \; n^{e_2-1} z_1^n;$$

$$\ldots$$

$$z_m^n, \; nz_m^n, \ldots, \; n^{e_m-1} z_1^n.$$

Remark. Note that if all the roots are distinct, then $m = k$, and any sequence obeying the recurrence relation is a linear combination of z_1^n, z_2^n, ..., z_k^n.

Remark. The specific linear combination is determined by the first k values of the sequence. The only difficulty is that the characteristic polynomial might not have explicit roots; for example $X^3 - X - 1$ compared to

$$X^3 - 2X^2 + X = X(X-1)^2.$$

We can see how this knowledge can help us solve recurrence relations in the following example:

Example 136. Consider $a_0 = -3, a_1 = 7$ with $a_n = 2a_{n-1} + 3a_{n-2}$ for $n \geq 2$. Find a closed form for a_n.

Solution. We can substitute into our recurrence

$$a_n = 2a_{n-1} + 3a_{n-2}$$
$$c \cdot r^n = 2 \cdot c \cdot r^{n-1} + 3 \cdot c \cdot r^{n-2}$$

This implies

$$c \cdot r^n - 2 \cdot c \cdot r^{n-1} - 3 \cdot c \cdot r^{n-2} = 0$$
$$r^n - 2r^{n-1} - 3r^{n-2} = 0 \quad \text{(dividing by } c\text{)}$$
$$r^2 - 2r - 3 = 0 \quad \text{(dividing by } r^{n-2}\text{)}$$
$$(r-3)(r+1) = 0$$

This tells us that the roots of our characteristic polynomial are 3 and -1. Thus our recurrence is solved by $a_n = c_1 3^n$ and $a_n = c_2(-1)^n$ for constants c_1, c_2. This gives us a general solution of

$$a_n = c_1 3^n + c_2(-1)^n.$$

Notice that up to this point we haven't dealt with our initial conditions. These initial conditions allow us to determine the specific values of c_1 and c_2 we want. We note that

$$-3 = a_0 = c_1 3^0 + c_2(-1)^0 = c_1 + c_2$$
$$7 = a_1 = c_1 3^1 + c_2(-1)^1 = 3c_1 - c_2.$$

This gives us a system of linear equations which we can solve to obtain $c_1 = 1$ and $c_2 = -4$. Thus our closed form for a_n is

$$a_n = 3^n - 4(-1)^n. \qquad \square$$

For inhomogeneous constant coefficients recurrence relations we have the following

Theorem. *If the sequence a_n satisfies the recurrence relation*

$$a_n = c_1 a_{n-1} + c_2 a_{n-2} + \ldots + c_k a_{n-k} + b_n,$$

then a_n can be written as $p_n + q_n$ where p_n satisfies the homogeneous recurrence relation $p_n = c_1 p_{n-1} + c_2 p_{n-2} + \ldots + c_k p_{n-k}$ and q_n is a particular solution.

Thus the only tricky part in solving inhomogeneous recurrence relations is finding a particular solution.

If b_n is polynomial in n of degree d, then we can find a particular solution which is polynomial of degree at most d in n. To find it, just plug it in the recurrence and identify the coefficients corresponding to the same powers of n and solve the linear system that arises. If b_n is an exponential function and the base is not any of the roots of the characteristic polynomial, then we can find a particular solutions which is exponential function in n with the same base.

Example 137. Let's jump back to our earlier example with $a_0 = -3, a_1 = 7$, and $a_n = 2a_{n-1} + 3a_{n-2}$. Let's change this recurrence instead to

$$a_n = 2a_{n-1} + 3a_{n-2} + n - 5.$$

Now what is our solution?

Solution. The first thing to note is that we found our homogeneous solution ("p_n") earlier, namely $p_n = c_1 3^n + c_2(-1)^n$. Now we just need to find a particular solution. Since our nonhomogeneous term is a linear polynomial, we'll guess a solution of the form $bn + c$. Then we have

$$bn + c = 2[b(n-1) + c] + 3[b(n-2) + c] + n - 5 = (5b+1)n - 8b + 5c - 5.$$

This tells us that $b = 5b + 1$ and $c = -8b + 5c - 5$. Solving these yields $b = -\frac{1}{4}$ and $c = \frac{3}{4}$. Thus our particular solution is $q_n = -\frac{1}{4}n + \frac{3}{4}$.

Finally we use our initial conditions to solve for the constants c_1 and c_2. We have

$$a_0 = -3 = c_1 + c_2 + \frac{3}{4} \qquad a_1 = 7 = 3c_1 - c_2 + \frac{1}{2}$$

Solving yields $c_1 = \frac{11}{16}, c_2 = -\frac{71}{16}$. Thus, our overall solution is

$$a_n = \frac{1}{16}(11 \cdot 3^n - 71 \cdot (-1)^n - 4n + 3). \qquad \square$$

As another example, if we want to solve the recurrence

$$a_n = 3a_{n-1} + 3a_{n-2} + 4^{n-2},$$

since the characteristic polynomial is $X^2 - 3X - 3$ and the base of $b_n = 4^{n-2}$ is 4, we are looking for a particular solution of type $a_n = 4^{n+A}$. Thus we have to solve, after simplification of 4^{n-2} from both sides, $4^{A+2} = 3 \cdot 4^{A+1} + 3 \cdot 4^A + 1$. Obviously this gives the solution $A = 0$ and thus a particular solution is $a_n = 4^n$.

Since we do not want to turn this appendix in to a list of rules, the two cases above should suffice for the scope of this book. Let us also remark that solving recurrence relations with constant coefficients, either homogeneous or inhomogeneous, can be done also using the the generating function attached to the sequence, $f(X) = \sum_{n \geq 0} a_n X^n$. An example of this method is the beginning example of generating functions chapter.

Glossary

Addition Rule. see Rule of Sum.

Adjacent. We say two vertices u, v of a graph are adjacent if there is an edge between them.

Binomial Coefficient. The binomial coefficients are the numbers

$$\binom{n}{k} = \frac{n!}{k!(n-k)!}.$$

Cardinality. The *cardinality* or *size* of a set A (denoted $|A|$) is the number of elements in that set.

Chromatic Number. The chromatic number of a graph G, denoted $\chi(G)$, is the minimum k such that G has a proper k-coloring.

Chromatic Polynomial. For all positive integers k, the chromatic polynomial of a graph G (denoted $\chi(G; k)$ is the function of k that counts the number of distinct colorings of G using k or fewer colors.

Combination. A combination is a subset of a group of objects. The number of combinations of k items from a set with a total of n distinct items is

$$\binom{n}{k} = \frac{n!}{k!(n-k)!}.$$

Complement of a set. If we have a *universal set* U which contains all of the objects we are interested in, we can define the complement of a set A (denoted A^c) as the collection of elements not in A (i.e., $A^c = U \backslash A$). For example, if we are working with the natural numbers, the complement of the set of even numbers would be the set of odd numbers. (*Note:* we have to have some universal set in order for the idea of a complement to make sense!)

Complete Graph. The complete graph on n vertices, denoted K_n, is the graph on n vertices containing all $\binom{n}{2}$ possible edges.

Convex. A polygon is said to be convex if all of its interior angles measure less than 180 degrees.

Degree. The degree of a vertex v in a graph is the number of edges of which v is an endpoint.

Derangement. A derangement of a set of objects is a permutation of those objects such that no object ends up in its orignal spot.

Disjoint. We say two sets A and B are *disjoint* if they have no elements in common (i.e. $A \cap B = \emptyset$).

Element. The notation $x \in A$ (read "x is in A" or "x is an element of A") says that the element x is included in the set A. On the other hand, we use the notation $y \notin A$ (read "y is not in A" or "y is not an element of A") to indicate that y is not included in the set A.

Empty Set. The *empty set* is the set which contains no elements. We denote it as $\{\ \}$ or $\emptyset$.

Expected Value. The expected value of a random variable X is given by

$$E[X] = \sum_{s \in S} X(s) \cdot P(s) = \sum_x xP(X = x),$$

where in the second sum x ranges over all possible values of X.

Generating Function. The generating function associated with a sequence $\{a_n\}_{i=1}^{\infty}$ is a formal power series of one variable. In particular, we have

$$f(x) = \sum_n a_n x^n.$$

Graph. A *graph* $G = (V, E)$ is a pair such that V is a finite set of vertices and E is a set of *edges* (i.e. 2-element subsets of V).

Intersection. The *intersection* of A and B (denoted $A \cap B$) is the set of all elements in both A and B: $\{x \mid x \in A \text{ and } x \in B\}$.

Invariant. An invariant is a property of an object or set of objects that does not change when certain transformations are applied to the object(s).

k-colorable. We say a graph is k-colorable if it has a proper k-coloring.

Leaf. A vertex of a tree that has degree 1 is called a leaf.

Multinomial Coefficient. A multinomial coefficient is of the form

$$\binom{n}{k_1, k_2, \ldots, k_m} = \frac{n!}{k_1! k_2! \cdots k_m!}$$

where $n = k_1 + k_2 + \cdots + k_m$.

Multiplication Rule. see Rule of Product.

Permutation. A permutation of an ordered list of distinct objects is an ordered list with the same objects, but possibly in a different order. The number of permutations of k distinct items chosen from a set with a total of n distinct items is $\dfrac{n!}{(n-k)!}$.

Probability. Given a set S, a probability is a function $P : S \to [0,1]$ such that
$$\sum_{s \in S} P(s) = 1.$$
For $A \subseteq S$, the probability of the event A is given by $\sum_{s \in A} P(s)$.

Proper k-Coloring. For $k \geq 1$ we say a graph $G = (V, E)$ has a *proper k-coloring* if there exists a mapping $f : V \to \{1, 2, \ldots, k\}$ such that $uv \in E$ implies $f(u) \neq f(v)$.

Random Variable. A random variable is a function defined on the sample space of a probability.

Recurrence Relation. A recurrence relation is an equation that defines elements of a sequence as a function of earlier elements in that sequence.

Rule of Product. If we have a sequence of n choices to make with X_1 possibilities for the first choice, X_2 possibilities for the second choice, and so on up to X_n choices for the nth choice, there are a total of $X_1 \cdot X_2 \cdots X_n$ ways to make our choices.

Rule of Sum. If $A_1, A_2, \ldots, A_n$ are pairwise disjoint sets (i.e., if no pair of sets have any elements in common), then
$$|A_1 \cup A_2 \cup \cdots \cup A_n| = |A_1| + |A_2| + \cdots + |A_n|.$$

Sequence. A sequence is an ordered list of elements. There may be infinitely many terms in a sequence.

Set. A *set* is a collection of different elements whose order is not important. We can specify a set by listing its elements such as $\{1, 2, 4, 8, 16\}$ or $\{3, 5, 7, \ldots, 19\}$. We can also use *set builder* notation where we specify a condition used to determine which elements belong to the set such as $\{x \mid 1 < x < 17\}$ or $\{(x, y) \mid y = 3x + 4\}$.

Set Difference. The *set difference* of the set A and the set B (denoted $A \backslash B$) is the set of elements that are in A but not in B.

Set Equality. Two sets A and B are *equal* (denoted $A = B$) if they contain the exact same elements. (*Note:* One common way to prove $A = B$ is to show that $A \subseteq B$ and $B \subseteq A$. Keep this in mind!)

Subgraph. A subgraph H of a graph G is a graph such that $V(H) \subseteq V(G)$ and $E(H) \subseteq E(G)$.

Subset. We say that a set A is a *subset* of a set B (denoted $A \subseteq B$) if every element of A is an element of B (i.e. $x \in A$ implies $x \in B$).

Tree. A tree is a graph that is connected and acyclic.

Union. The *union* of two sets A and B (denoted $A \cup B$) is the set of all elements in either A or B: $\{x \mid x \in A \text{ or } x \in B\}$.